Quantifiers in Action
Generalized Quantification in Query, Logical and Natural Languages

ADVANCES IN DATABASE SYSTEMS
Volume 37

Series Editors

Ahmed K. Elmagarmid
Purdue University
West Lafayette, IN 47907

Amit P. Sheth
Wright State University
Dayton, Ohio 45435

For other titles published in this series, go to
www.springer.com/series/5573

Quantifiers in Action
Generalized Quantification in Query, Logical and Natural Languages

by

Antonio Badia
Computer Engineering and Computer Science Department
Speed School of Engineering
University of Louisville
Louisville, KY, USA

 Springer

Author:
Antonio Badia
Computer Engineering and Computer Science Department
JB. Speed School of Engineering
University of Louisville
Louisville, KY 40292 USA
abadia@louisville.edu

Library of Congress Control Number: 2009921815

ISBN: 978-0-387-09563-9   e-ISBN: 978-0-387-09564-6

Printed on acid-free paper

9 8 7 6 5 4 3 2 1

springer.com

To Mindy, Emma and Gabi: I'm finally done!

Preface

The goal of this book is to help conquer what this author sees as dangerous divides in the study of databases in general, and of query languages in particular: the divide between theory and practice, and the divide between purely formal approaches and those with a wider concept of communication.

This is, of course, a very tall order, and there is no claim here that the mission has been accomplished. We live in an era of specialists (people who know a lot about a little), and the explosion of available material in any given field makes it quite difficult even to keep current with one's area, never mind exploring a different one. Presenting all the material this book touches upon in depth is nearly impossible. Rather, the book introduces the most basic results of different fields. It is then likely that an expert in any given area (logic, databases, or formal linguistics) will find that the material discussed here is very elementary. This is part of the price to pay when trying to reach the widest audience. The book is intended just to present the beginning of an exploration, and its aim is to encourage others to pursue research further. Necessarily, the coverage is limited and subjective, but hopefully it will become an *entry point* for the interested reader, who will be encouraged to delve further into the subject using some of the bibliography provided.

It may seem strange to pick a a technical, specialized topic like Generalized Quantification for this kind of endeavor. However, Generalized Quantifiers are a perfect example of a practical theory. An idea that started in the seminal papers by Mostowski ([76]) and Lindstrom ([70]) as a theoretical investigation into a new concept found unsuspected applications first in linguistics and then in query languages. Generalized Quantifiers motivate insights into optimization of set-based queries, Question Answering, and Cooperative Query Answering. This book tries to gather all those insights together. As such, the book tries to walk a line between theory and practice. There is out there a large collection of work on GQs; however, they all tend towards the purely theoretical side ([4]) or towards a particular application, usually linguistics ([94]). The book strives the cover the middle ground.

Depending on their point of view, readers may feel that we have gone too far to the theory side or to the application side. Others still may feel that we have not gone far enough. There are no doubt many other possible applications of the concept, and much more is known right now that we can hope to include in our first chapters. However, by bringing together all this material we hope to make people aware of what others are doing. If we happen to encourage collaborations and help establish links where before there were none, then we consider that we have done our job.

Before getting into the subject, it is time to give credit where credit is due. This book represents a summary of research carried out by the author over the past years. In particular, section 4.3 of chapter 4 is based on [10]; sections 6.2, 6.3 and 6.4 of chapter 6 are based on [13]; chapter 7 is based on [9] and chapter 8 is an update of work reported in [11]. Along the way, there were many people who helped carry out the research, and this is the opportunity to thank them, as they made this book possible. This list should certainly include the members of my doctoral dissertation committee (Dirk van Gucht, Ed Robertson, David Leake and Larry Moss), as it all got started there. The collaboration with Stijn Vansummeren was a pleasure; I learned much from it. My graduate students contributed considerably to this effort; special mention should go to Bin Cao, Brandon Debes and Dazhuo Li for their work in different parts of the project. Bin helped develop the interpreter introduced in chapter 5; Dazhuo worked an approach to linear prefixes, shown in chapter 6; and Brandon implemented a system that transformed our project from a paper-and-pencil approach to executable software[1]. This material is based upon work supported by the National Science Foundation under Grant No. IIS-0347555, and therefore special mention goes to the National Science Foundation, which generously supported the research described here under a CAREER Award, and to my two program managers, Le Gruenwald and Maria Zemankova, who exhibited a tremendous amount of patience with me -I am very indebted to both of them.

Finally, I cannot honestly say that Mindy, Emma or Gabi's contributions have been technical. In fact, I cannot say that repeatedly asking "Tickle me, papa!" is a contribution at all. But I know that without them, I would have never spent the time and energy I did on this, or any other project -it would just not seem worthwhile.

Louisville, KY   *Antonio Badia*
                                                              December 2008

---

[1] The result of his efforts is accessible at http://qlgq.spd.louisville.edu/index.php.

Contents

1

Introduction

This monograph is written with two purposes: to help bridge the gap between theory and practice in databases, and to help bridge the gap between research in query language and research in other areas (outside databases, and some outside Computer Science) that are clearly related to querying. The motivation behind the book, then, is a belief that the gaps exist, and that this is a bad situation.

Databases are a strange area. On the one hand, they are clearly an *applied* field. Databases (and related services) are a multi-billion, world-wide software industry. Some advances (for instance, in query optimization) make it to market in relatively short time. Changes in the industry (for instance, the move to *data warehouses*) provoke comparable changes in the field, creating whole subareas of research. On the other hand, databases are a *theoretical* field. Since the definition of the relational model by Codd, its main concepts have been tied to logic ideas, and logician's methods have been used to study the model abstractly. This study has blossomed in entirely new subareas, like *descriptive complexity* ([56]) and *finite model theory* ([27, 69]). Thus, databases is, at the same time, a clearly applied and a clearly theoretical field of study.

This would be a good thing if theory and practice walked hand in hand. However, this does not always happen. To quote a familiar dictum, *"In theory, theory and practice are the same. In practice, they are not."*[1]. While occasionally theory occupies itself with questions of practical interest, and practice does generate questions of theoretical interest, the mainstream body of theory and practice remain separated. This is not due to the will of the researchers -many theoreticians find delight in seeing theoretical developments put to good use in practice; and they are also eager to apply their skills to problems motivated by new areas of practice. It is also the case, in my opinion, that many practitioners consider the establishment of a solid, formal foundation for a practical problem a positive development. In fact, most database

---

[1] This saying, as well as the next one, have been attributed to a variety of people, so they will go uncredited here.

A.Badia, *Quantifiers in Action: Generalized Quantification in Query. Logical and Natural Languages*, Advances in Database Systems 37, DOI: 10.1007/978-0-387-09564-6_1,
© Springer Science+Business Media, LLC 2009

people (certainly this author) adhere to another familiar dictum, *"There is nothing as practical as a good theory"*. But theory and practice have different goals, demand different approaches and produce different results. Most of the time, the disconnect is present. It is for this reason that it is a pleasure to work on a subject that gives an opportunity to bridge this gap.

As for our second purpose, it is not difficult to realize, after a bit of thought, that querying is, first and foremost, a *linguistic* activity: queries are posed in a language. However, there are two levels at which this activity takes place. First, a cognitive agent has to come up with a question. Such agents are, usually, human beings, and the questions they come up with are expressed in natural language. Only when the necessity to pose the question to a computer arises there is a corresponding need to translate the question into a formal language. Let us agree to reserve the term *question* to refer to expressions in a natural language, and *query* to refer to expressions in a formal language. The database researcher (and the theoretical computer scientist) deals with queries *only*. The linguist (and the philosopher) deals with questions *only*. But surely a link is missing. In real life, queries start as questions. Making sure a query correctly reflects the intended question is important. Learning which kinds of questions (as opposed to queries) are most useful (or common, or important, or whatever other measure one wants to use) is also important. This importance has been reflected in the fast development, in the last few years, of the field of Question Answering, in which users pose questions to the computer directly in natural language, and hence it is the software that must come up with an appropriate translation from question to query ([99]). Information Retrieval, also prevalent nowadays as a research field due to the explosion of the World Wide Web, straddles a middle line between queries and questions -at least, that is how this author regards keyword search ([14, 18, 73]). It is perhaps time, then, to take the relationship between questions and queries seriously.

Once our purpose has been explained, the next question is: why use Generalized Quantification as the tool to develop a research plan for this purpose?

Generalized Quantification started as a narrow concept applied to a particular problem and then, like many good ideas, took off with a life of its own. The great logician Mostowski considered the limitations of expressive power of first order logic (henceforth, FOL) and focused on one that seemed somewhat troublesome to him: the inability of FOL to distinguish between finite and infinite sets ([76]). The concept of infinity being so important for the foundations of mathematics, he considered that FOL had to be able to capture the notion (there were contrary opinions: for other people, the notions of finite and infinite sets were not purely logical notions and hence did not belong in FOL). So he considered how to add to FOL the ability of expressing infinity. The challenge was doing so in a manner that was *minimal,* in the sense that minimal changes were introduced to both the syntax and the semantics of the language. He introduced a new symbol, $\mathbf{Q}_{\aleph}$, and used in formulas as follows:

$$Q_{\aleph} \ x \ \phi(x)$$

where ϕ was an arbitrary FOL formula with x the only free variable. The interpretation of this formula is, as usual, given in terms of whether an arbitrary model $\mathcal{M}$ would satisfy it[2]. Given the set of values a from the universe of $\mathcal{M}$ such that $\mathcal{M}$ satisfies $\phi(a)$ (that is, when a is substituted by x, we get a sentence true in $\mathcal{M}$), Mostowski stipulated that if such a set is infinite numerable, then the sentence above is true in $\mathcal{M}$.

The subtle but important idea behind that notion is that, by introducing a new logical operator that captures exactly the notion we intended to capture ("infiniteness"), we are adding to the language *only* the necessary machinery to deal with the new concept. By studying the resulting language, one could learn what the concept exactly entails. For instance, if some property can also be expressed in the extended language that could not be expressed in plain *FOL*, that such property must be somehow related to infinity. A trivial example is that of finiteness, which can be expressed in the language simply by using negation on the formula above.

A few years later, as part of its meta-logic studies on exactly what constitutes a logic (and on what makes *FOL* such an important language among many possible) Lindstrom revisited the issue of Generalized Quantification and realized that it could be put in a more general setting ([70]). Just like he was doing for other logic concepts, Lindstrom formally defined the idea of Generalized Quantification, of which Mostowski's quantifier was only one example[3].

The idea then became part of the logician's tool and was studied by researchers; however, it didn't make any impact outside the logical community. But this all changed with time. Nowadays, the concept of Generalized Quantification is heavily used in at least two communities outside its original setting: theoretical Computer Science and Formal Linguistics.

Theoretical Computer Science's main theme is the study of computational complexity, which classifies problems according to the resources (time and space) that any algorithm needs to compute a solution to the problem. Since there can be many algorithms for a given problem, careful study of the problem is needed to establish such results -which should hold for all possible algorithms that compute solutions to the problem. In [30], Fagin introduced the idea of using some logic language to describe the problem, and studying the properties of such logic in order to determine properties of the problem. This was a significant new viewpoint, since logics are *declarative* languages: they simply allows us to specify *what* the problem is, not *how* a solution looks like (which is what algorithms do). Thus, the study linked issues of computational complexity with issues of expressive power in logic, and a large number of results followed, starting a subfield that is usually called *declara-*

---

[2] For readers not acquainted with logic, chapter 2 introduces all needed basic notions.

[3] Lindstrom's definition is given in chapter 3.

tive complexity. The explosion of results was facilitated, in part, by the body of knowledge that logicians had developed over time.

The idea had a direct impact on query languages. It is well known that relational query languages are simply versions (more or less disguised) of FOL. The limitations in expressive power of FOL have became an important issue and, since Generalized Quantifiers offer a way to overcome some of those limitations, logics with Generalized Quantifiers have been studied in depth in this setting. As its name indicates, Generalized Quantifiers are an extension of the idea of quantifier in FOL. There, the notion of quantifier is *fixed*: first-order quantifiers are a closed class, capturing only the most basic concept. Generalized Quantifiers consider the class open: you can add new quantifiers simply by defining them, subject to some very basic rules that make sure any quantifier behaves in a logical way (what this means will be defined in Chapter 3). It turns out that this is a natural way to capture properties that are not expressible in FOL; this makes Generalized Quantifiers a great tool to investigate expressive power, and to add it in a controlled manner to query languages.

Formal linguists attempt to give a description of the meaning of natural language statements in a formalized language. FOL is one of the popular choices for a target language. The inadequacies of such a choice have been known for a long time. In a work that touched most later research, Richard Montague argued that natural language could be successfully modeled with tools that were better suited for the task ([26]). However, he used heavy equipment: a logic derived from lambda calculus that contained many second-order constructs. Surely there had to be something between FOL and the sophisticated tools of Montague that was adequate for the analysis of at least some simple fragments of natural language. In 1981, in a seminal paper, Barwise and Cooper introduced the idea of using FOL extended with Generalized Quantifiers for the formal analysis of linguistic expressions ([16]). Using a suitably simplified notion of Generalized Quantification, Barwise and Cooper established a basic framework that has lasted until today and has originated its own significant body of research (see, for instance, [94]).

We have, then, a theoretical concept that can have practical application, and that has spanned several areas of research. The book's thesis is that Generalized Quantification is a good idea that can be profitable applied to practical matters of questioning and querying. Going back to the old saying that *there is nothing as practical as a good theory*, Generalized Quantification fits the bill perfectly. At the same time, applying the idea may not be easy. Using the concepts introduced here in a practical scenario involves being able to implement them efficiently (always a strong consideration in databases), and showing its usefulness for practical purposes. Thus, unlike past research that dealt with theoretical issues of complexity and expressive power, the aim here is to work with a suitably limited version of the concept but to show that such version can still be useful and can be implemented efficiently. This endeavor is helped by the inter-disciplinary approach mentioned above, as the inspiration of what version of the concept to use in queries comes from research on linguistics. And this is the other main goal here: to show how concepts and techniques developed in one field help illuminate challenges in others. The book starts with a formal view of queries and gradually widens the scope to incorporate more and more issues about questions. To develop this program, Chapter 2 gives some basic background in logic focused on the traditional notion of quantification. The idea is to make the rest of the book accessible to readers without a background in logic, making the book as self-contained as possible. Readers with an adequate background may skip this chapter, which only introduces elementary concepts. Chapter 3 introduces the notion of Generalized Quantifier (in fact, two definitions are given) and some basic properties. We have chosen to introduce further properties later, right when they are needed, so that their motivation is clear. Chapter 4 introduces the Query Language with Generalized Quantifiers ($QLGQ$), which provides us with a general setting in which to study issues of quantification without being distracted by this or that particular syntax. The language is a simple variation of Datalog-like notations, and hopefully most readers will have no trouble becoming familiar with it. Chapter 5 introduces what we consider the most common case of Generalized Quantification for practical use, and concentrates on giving an efficient implementation for this case. Chapter 6 studies the use and implementation of (generalized) quantifier prefixes. We will see that, even in the simplest of settings, some questions come up that may not be evident on first thought. Then chapter 7 introduces, following a more linguistic motivation, the use of Generalized Quantification to deal with *pragmatic* issues in querying, and chapter 8 finishes the change to a linguistic setting by showing the use of Generalized Quantifiers in Question Answering. Finally, chapter 9 sketches how Generalized Quantifiers can be used in other settings, for instance distributed computing, so prominent nowadays with the raise of peer-to-peer systems and cloud computing. While the research in this chapter is just in its beginning stages, it will hopefully be enough to show the adaptability and promise of the approach. We close the book with a short discussion in chapter 10.

2

Basic Concepts

2.1 From Propositional to First Order Logic

In this chapter we introduce some basic logic concepts, focusing on those ideas
related to quantification. This chapter is for readers with no background in
logic, and it only includes some core concepts that facilitate further reading.

Propositional (also called *zero-order*) logic is the basic building block of
First-Order Logic (recall that we abbreviate it as FOL). In fact, we have the
same basic logic connectives:

- if ψ, φ are formulas, then $\psi \wedge \varphi$ is a formula (conjunction);
- if ψ, φ are formulas, then $\psi \vee \varphi$ is a formula (disjunction);
- if ψ is a formula, then $\neg\psi$ is a formula (negation);

It is customary to also allow formulas of the form $\psi \rightarrow \varphi$ and $\psi \leftrightarrow \varphi$, but
they are considered just syntactic sugar, since $\psi \rightarrow \varphi$ is equivalent to $\neg\psi \vee \varphi$,
and $\psi \leftrightarrow \varphi$ is equivalent to $(\psi \rightarrow \varphi) \wedge (\varphi \rightarrow \psi)$. Because of the recursive
nature of the definition, arbitrarily complex formulas can be formed.

What truly distinguishes FOL from propositional logic is *quantification*.
Propositional logic allows only *propositions*, statements that are either true
or false: "yesterday it rained", "3 is greater than 5". In FOL, statements
are made about individuals, by stating relationships that may hold among
them. As an example, the statement "3 is greater than 5" would be expressed
in FOL with a formula like $> (\mathbf{3}, \mathbf{5})$, where the symbol '$>$' denotes a binary
relation. In propositional logic, such an statement would be considered atomic
(without parts), and denoted by a single symbol, like P. In FOL, statements
are no longer atomic. Since they are about individuals, though, the next thing
we need is some way to tell *which* individuals we are talking about. FOL uses
terms to denote individuals. Terms are of two basic types: *constants* or names
that denote a certain individual unequivocally, or *variables*, which stand for
an individual without denoting a particular one.

FOL languages then, will be made up of atomic formulas constituted by
relations and terms. Such formulas can then be combined with the Boolean

A.Badia, *Quantifiers in Action: Generalized Quantification in Query, Logical and Natural
Languages*, Advances in Database Systems 37, DOI: 10.1007/978-0-387-09564-6_2,
© Springer Science+Business Media, LLC 2009

connectives mentioned above. It is customary to assume an infinite set of *variable symbols*, usually denotes $x, y, z, x_1, \ldots$ (it is also convenient to assume an infinite set of *constant symbols*). For example, when talking about natural numbers, x, y denote arbitrary natural numbers, while the constant **5** denotes a particular natural number.

A *vocabulary* (also called *signature* in some logic books) L is a non-empty set of symbols, which are classified as: relational symbols, function symbols and constant symbols. Some signatures may have only functions (they are called *algebras*); some have only relations (they are called *relational*). We restrict ourselves to the relational, finite case here[1]. For each relational symbol R, it is customary to include a natural number called the *arity* of R (in symbols $\alpha(R)$) in the signature. This corresponds to how many elements R takes. Note: a relation of arity 1 will be called a *property* and is customarily identified with its underlying set (instead of with a set of 1-tuples). Relations of arity 2 are called binary, and so on.

Given a relational signature $L = \{R_1, \ldots, R_m\}$, an L-structure $\mathcal{A}$ is an tuple $< A, R_1, \ldots, R_m >$, where A is a non-empty set (called the *domain* or *universe* of $\mathcal{A}$) and R_i is a relation on $A^{\alpha(R_i)}$, for all $i = 1, \ldots, m$.

An *atomic formula* is an expression of the form $R(t_1, \ldots, t_n)$ where R is a relational symbol of arity n and t_i is a term ($i = 1, \ldots, n$). Note that the *infix* notation (the operator between the terms, instead of in front of them) is more customary. For instance, $\psi(x) = x > \mathbf{5}$ (instead of the more formal $> (x, \mathbf{5})$) is an atomic formula. Hence, our idea of atomic formulas includes formulas of the form $t_1 \theta t_2$, where t_1, t_2 are terms and θ is a comparison operator, since θ is considered a binary relation.

Note that we stated that propositions were always true or false. Formulas, on the other hand, may not always have a truth value. As an example, the formula $x > 5$ is, intuitively, true or false depending on what number x stands for. Different values for x will yield different truth values for the formula. *Interpretations* (sometimes called *assignments*) are functions from the set of variables to the domain of a structure, and are used to indicate which value a given variable is intended to take. With an assignment, then, a formula can be given a truth value; however, most of the interest is in cases where the assignment is really not relevant -as it is arbitrary (this is formally explained later). Quantification is a way to handle variables systematically and assign truth values to certain formulas.

2.2 Quantification

A *well formed formula (wff)* is an expression built up from atomic formulas and the rules for conjunction, disjunction, negation, and the following (quantifier introduction):

---

[1] Some vocabularies may be infinite, but we will not have a use for them in this book.

- if φ is a formula and x a variable, then $\exists x \varphi$ is a formula.
- if φ is a formula and x a variable, then $\forall x \varphi$ is a formula.

When there is a risk of ambiguity, parenthesis are used to make formulas non-ambiguous. In fact, one of the most important characteristics of formal languages (compared to natural languages) is that each formula in a formal language has a *unique reading*, i.e. ambiguity is not tolerated at all. In natural languages, ambiguity is a fact of life and is even necessary for some purposes (e.g. humor). For instance, $\psi_1 \wedge \psi_2 \vee \psi_3$ is written as $(\psi_1 \wedge \psi_2) \vee \psi_3$ or as $\psi_1 \wedge (\psi_2 \vee \psi_3)$, depending on what was intended[2].

We will always assume a special binary relation, *equality*, denoted by $=$. As is customary, we will write $a = b$ instead of the more formal $= (a, b)$.

Given variable x, wff ψ, we define that x *occurs free* in ψ:

- For atomic ψ x occurs free in ψ iff x occurs in ψ.
- x occurs free in $\neg \psi$ iff x occurs free in ψ.
- x occurs free in $\psi \wedge \varphi$ iff x occurs free in ψ or x occurs free in φ.
- x occurs free in $\forall y \; \psi$ iff x occurs free in ψ and $x \neq y$.

A variable that is not free is *bound*. Bound variables, is easy to see, are the ones tied to a quantifier. As an example, in $\psi(x) = x > \mathbf{5}$, x occurs free.

A *sentence* is a formula with no free variables. As an example, $\mathbf{6} > \mathbf{5}$ is a sentence since it has no variables (and therefore no free ones). As another example, $\forall x (x > \mathbf{5})$ is a sentence, as well as $\exists x (x > \mathbf{5})$. In both sentences, there is a variable, x, but it is bound.

2.2.1 Semantics

The basic semantic relationship is that between expressions in the language and structures. A L-structure $\mathcal{A}$ gives meaning to the symbols in a language with vocabulary L, in the sense that for any relational symbol R, the structure provides a relation $R^{\mathcal{A}} \subseteq A^{\alpha(R)}$. Also, the structure provides a meaning for function symbols and constants in L -each constant is related to some element in the domain.

A sentence ψ of a language with signature L will be either true or false in a L-structure $\mathcal{A}$. If it is true, we say that $\mathcal{A}$ *is a model of* ψ (in symbols, $\mathcal{A} \models \psi$). In order to define this formally, we need to deal with the variables, since a formula with free variables is neither true nor false until a value has been assigned to the variables. Let $s : V \to A$, a function from the set of all variables to the universe of $\mathcal{A}$. This function is what we called an *interpretation*. We allow this function to extend to constants by simply making it aware of the reference of constants in the structure. Given interpretation s, variable x, element $a \in A$, $s(x/a)$ is a function that is exactly like s except that it gives x the value a; that is, for every variable y:

---

[2] Another standard way to get rid of ambiguity is to give connectives a rank or preference, so that in case of conflict one reading is always preferred.

$$s(x/a)(y) = \begin{cases} a & \text{if } x = y \\ s(y) & \text{if } x \neq y \end{cases}$$

We define $\mathcal{A} \models \psi[s]$, (read: $\mathcal{A}$ satisfies ψ with s, or $\mathcal{A}$ is a model of ψ under s) as follows:

- Atomic formulas: $\mathcal{A} \models R(t_1, \ldots, t_n)[s]$ iff $(s(t_1), \ldots, s(t_n)) \in R^{\mathcal{A}}$.
- Negation: $\mathcal{A} \models \neg\psi[s]$ iff $\mathcal{A} \not\models \psi[s]$
- Disjunction: $\mathcal{A} \models (\psi \vee \varphi)[s]$ iff $\mathcal{A} \models \psi[s]$ or $\mathcal{A} \models \varphi[s]$.
- Conjunction: $\mathcal{A} \models (\psi \wedge \varphi)[s]$ iff $\mathcal{A} \models \psi[s]$ and $\mathcal{A} \models \varphi[s]$.
- Quantification:
  - $\mathcal{A} \models \exists x \ \psi(x)[s]$ iff there exists $a \in A$ such that $\mathcal{A} \models \psi[s(x/a)]$.
  - $\mathcal{A} \models \forall x \ \psi(x)[s]$ iff for every $a \in A$, $\mathcal{A} \models \psi[s(x/a)]$.

We give a few examples, to show the generality and power of this seemingly simple definition. Let $\mathcal{G} = (V, N)$ where $V \neq \emptyset$ and $N \subseteq V^2$. Thus, G is a graph. The following formulas are examples on the vocabulary with one binary symbol, N:

- $\forall x \forall y (N(x,y) \leftrightarrow N(y,x))$ is true if the graph is undirected. Note that this is a sentence.
- $\forall x \ \forall y \ N(x,y)$ is true if the graph is a clique. This is a sentence too.
- $\varphi(x,z) = \exists z \ N(x,y) \wedge N(y,z)$ is true in structures with at least two nodes connected by an edge (note that $s(x) = s(z)$ is possible!). That is, there is a path of length at most two between x and y. Note that this is a formula. To determine whether it is true or not in a given graph (model), we need to fix two nodes (or one node, when $s(x) = s(z)$) in the graph.

As another, more mundane, example, assume a universe of people and a binary relation $L(x,y)$, intended to mean x *loves* y. Then the sentence

$$\forall x \forall y L(x,y) \rightarrow L(y,x)$$

states that all loves are corresponded (whenever x loves y, y loves x in return). A happy world indeed. In contrast, the sentence

$$\exists y \forall x \ \neg L(x,y)$$

states that there is someone who is not loved by anyone -a sad soul.

Several things are important enough about the crucial definition above to warrant closer attention. First, the interpretations are infinite, since the set of variables is infinite. However, in evaluating a formula all it matters is what happens to the variables in the formula (and the number of variables in a formula is finite):

Lemma 2.1. *Let s_1, s_2 be two interpretations into A that agree on in all the variables that occur free in wff ψ. Then $\mathcal{A} \models \psi[s_1]$ iff $\mathcal{A} \models \psi[s_1]$.*

In fact, since interpretations only take care of free variables, in a sentence (where there are none), the interpretation is irrelevant to establishing meaning:

Lemma 2.2. *For a sentence α, model $\mathcal{A}$, either $\mathcal{A}$ satisfies α under every possible interpretation, or $\mathcal{A}$ does not satisfy α under any interpretation.*

If the former, we say that α is true in $\mathcal{A}$ ($\mathcal{A} \models \alpha$).

The final point concerns the quantifiers, our main object of study. First, note that the universal quantifier, appealing to its name, means basically that a property holds of all the objects in the universe (the domain under consideration), while the existential quantifier (also appealing to its name) means that there is *at least one* object in the universe with the property (if there are more than one, that's no problem!). So the bound variables in a formula are not evaluated by giving them a value through the interpretation; rather, they have to take all values (for universal) or pick one certain value (for existential). That is why these variables are not considered free; it is also the reason for Lemma 2.2. Several consequences of the definition that are important are listed next:

- For any formula $\forall x \, \psi$, structure $\mathcal{A}$, new variable y (i.e. a variable not appearing in ψ), $\mathcal{A} \models \forall x \, \psi(x)[s]$ iff $\mathcal{A} \models \forall y \, \psi(y)[s]$, and $\mathcal{A} \models \exists x \, \psi(x)[s]$ iff $\mathcal{A} \models \exists y \, \psi(y)[s]$. That is, bound variables are just labels to relate quantifiers to certain positions in formulas. Changing a bound variable does not affect a formula's meaning.
- $\forall x \psi(x)$ is equivalent to $\neg \exists x \neg \psi(x)$. In English, this says that "every thing has ψ" and "there is no thing that does not have ψ" are equivalent. You can see that this is the case by reasoning from the rules above.

For a sentence σ, $Mod(\sigma)$ is the class of all structures that are models of σ. The above results state that $Mod(\psi)$ is *closed under isomorphism*, that is, if $\mathcal{A} \in Mod(\psi)$ and $\mathcal{A} \cong \mathcal{B}$, then $\mathcal{B} \in Mod(\psi)$.

This is typical of classes defined logically. Logics talk about the structure of a model, not about the particular components -although by using constants, they may mention them, but all they can do is talk about the properties of such and such element. And if the structures are isomorphic, for each element in one there is an element in the other with identical properties. In fact, on deciding *what* can be called a logic, this is a fundamental condition on the definition, called the *Isomorphism Condition*, that isomorphic structures cannot be distinguished by formulas of the logic.

2.2.2 Meaning

What is the *meaning* of a sentence ψ? Any sentence s can be classified as in one of three classes:

- s is true in all models ($Mod(s)$ is the class of all models); sometimes such sentences are called *tautologies*.
- s is never true ($Mod(s) = \emptyset$); sometimes such sentences are called *contradictions*.
- s is true in some models and false in some others.

The last case is the more interesting one. Intuitively (and also informally), we say that the meaning of σ is whatever all structures in $Mod(\sigma)$ have in common. This is not exactly correct; there may be several things that all models in $Mod(\sigma)$ have in common, but is a rough approximation. As an example, let $\sigma = \exists x \exists y (x \neq y)$. Then all models of σ have one thing in common: there are at least two elements in their domain. There may be a lot of other stuff going on on those structures, but this is one thing that they all are guaranteed to have in common. Thus, we say that σ means "there are at least two elements".

2.3 More on Quantification

Here we discuss some properties of FOL quantifiers that give more insight into their meaning and usage.

2.3.1 Quantifier Scope and Prefixes

FOL formulas are read "from left to right". In some cases, this makes no difference, as operators are symmetric (for instance, for conjunction or disjunction, the order does not matter: $\varphi \wedge \psi$ and $\psi \wedge \varphi$ are equivalent).

As seen in some of our previous examples, a formula may contain several quantifiers. When quantifiers appear in a formula, they precede (are to the left of) the part of the formula where the variable they bound is used. Thus, we write $\forall x \varphi(x)$ and not $\varphi(x) \forall x$. However, individual quantifiers may appear "intermixed" with other formulas, as in $\forall x (\varphi(x) \wedge \exists y \psi(y))$; other times, several quantifiers may appear together before a formula: $\forall x \exists y \forall x \varphi(x, y, z)$. In such a case, we talk of a quantifier *prefix*; in the above example, $\forall x \exists y \forall x$ is the prefix. But the only requirement is that a quantifier binding a variable appears before (to the left of) the first mention of that variable.

We saw that for some connectives, order does not matter. when quantifiers are involved, order matters a lot: $\forall x \exists y$ is not the same as $\exists y \forall x$. As an example, assume our past example of people and the binary relation L (loves). Then, the sentence

$$\forall x \exists y \; L(x, y)$$

states that, for any individual in the domain, there is someone that such individual loves; in English, "everybody loves somebody". But the sentence

$$\exists y \forall x \; L(x, y)$$

states that a certain individual is such that all elements in the domain love that individual; in English, "someone is loved by everyone". We would say that, in real life, the first sentence is much more likely to be true than the second one. The *scope* of a quantifier is the part of the formula which it affects, and is all the formula to the immediate right of the quantifier. The scope of a quantifier works left-to-right, the same order in which we read. Thus, the quantifier we read first takes precedence over a quantifier that we read later. However, order among universal quantifiers is irrelevant (ie. $\forall x \forall y$ is the same as $\forall y \forall x$) and the order among existential quantifiers is irrelevant (ie. $\exists x \exists y$ is the same as $\exists y \exists x$), but the order among universal and existential quantifiers is essential.

Whenever quantifiers are dispersed around in a sentence (or formula) the quantifiers can be moved "to the left" until they form a prefix, since they can pass over variables that they do not bound. For instance, $\forall x (R(x) \wedge \exists y P(x, y))$ can be written as $\forall x \exists y (R(x) \wedge P(x, y))$. If there is a conflict because one variable name is used in two quantifiers, we can always "rename" the variable to avoid conflicts. So

$\forall x \exists y (R(x, y) \rightarrow \exists x S(x))$

can be transformed into the equivalent

$\forall x \exists y (R(x, y) \rightarrow \exists z S(z))$

and then written as

$\forall x \exists y \exists z (R(x, y) \rightarrow S(z))$

Note the similarity with programming language variables, when the programming language has block scope: a variable has as scope the block in which it was defined; if there is a name conflict with an inner block, a *hole* is created in the scope. Further, the variable can be changed without affecting the semantics of the program.

Note that, while quantifiers can be moved in front of a formula *without* quantifiers, we cannot move a quantifier in front of another, since quantifiers order does matter. Thus, when putting a formula in PNF, we always move first the leftmost quantifier, and proceed in strict order.

There are two consequences of the facts just mentioned. First, every *FOL* sentence can be written in *prenex normal form* (PNF): all quantifiers in front, forming a prefix, followed by a quantifier-free sentence. So the sentence has the form

$$Q_1 x_1 \ldots Q_n x_n \varphi$$

where Q_i is one of $\forall$, $\exists$, for $1 \leq i \leq n$ and φ is a quantifier-free formula. If Q_1 is $\exists$, we call the formula *existential*; if $Q_1 = \forall$ we call the sentence *universal*. The following definitions make this formal.

Definition 2.3. *A formula is in* prenex normal form (PNF) *if it is in the form $Q_1 x_1 \ldots Q_n x_n \varphi(x_1, \ldots, x_n)$, with φ being quantifier free.*

Definition 2.4. *A sentence ψ is universal if it can be written in PNF using only universal quantifiers. A sentence ψ is existential if it can be written in PNF using only existential quantifiers.*

The second consequence is the following: when we write a sentence in PNF, we can divide the quantifier prefix into *blocks* as follows: a block is a sequence of adjacent quantifiers of the same type. So, one or more existential (respectively, universal) quantifiers that follow each other form a existential (respectively, universal) block. As an example the following sentence has its blocks marked:

$$\underbrace{\forall x \forall y}_{1^{st} block} \quad \underbrace{\exists z \exists w}_{2^{nd} block} \quad \underbrace{\forall u}_{3^{rd} block} \quad \underbrace{\exists v \exists t}_{4^{th} block} \quad \varphi(x, y, z, w, u, v, t)$$

The even numbered blocks are existential, the odd-numbered ones, universal. The third block consists of only one quantifier.

The observations that we made before can be rephrased as follows: it is ok to move quantifiers *within* the block they belong to, but not across blocks. In fact, old results from [100] and KW2 state that any sentence obtained from a given one by moving quantifiers within a block is equivalent to the original one, while any sentence obtained by moving quantifiers across blocks is different. In particular, [63] proved the *Linear Prefix Theorem*, stating that for each two different quantifier prefixes Q_1 and Q_2 of the same length there is a Q_1 sentence (i.e. a sentence made up of Q_1 as a prefix) that is not equivalent to any Q_2 sentence, while [100] shows that in the case of first order logic prefixes are equivalent (in the sense of semantically indistinguishable) iff they are literally the same, i.e. they only differ in variables, but order and type of quantifier are preserved.

2.3.2 Skolemization

To see the dependence between different quantifiers, the concept of *Skolem function* comes in handy. A process called *Skolemization* is carried out as follows: let φ be a formula in PNF, that is, $\varphi = Q_1 x_1, \dots Q_n x_n \psi$ as before. Then we define the following sequence of formulas:

1. $\varphi_0 = \varphi$;
2. $\varphi_i =$ the sentence obtained by the following procedure:
   a) if $Q_i = \forall$: remove $Q_i x_i$ from the beginning of φ_{i-1}.
   b) if $Q_i = \exists$: remove $Q_i x_i$ from the beginning of φ_{i-1} and replace all occurrences of x_i by $f(\bar{y})$ where f is a new function symbol and $\bar{y} = Fvar(\varphi_{i-1})$.

The function symbols introduced in this process are called *Skolem functions*. Clearly the process ends at φ_n; which is called the Skolemization of φ. Skolemization produces a quantifier-free formula, where the complexity (arity) of the terms created by functionals shows the original quantifier prefix. For instance, formula $\forall x \exists y \forall z \exists u R(x, y, z, u)$ becomes $R(x, f_1(x), z, f_2(x, z))$.

Let, for formula φ, $sk(\varphi)$ denote its Skolemization. Then let $\bar{f}$ be all function names appearing in $sk(\varphi)$ and $bar x$ all free variables appearing in $sk(\varphi)$. The formula

$\exists \bar{f} \forall \bar{x} sk(\varphi)$

is called the *quantified Skolemization* of φ, in symbols $qsk(\varphi)$[3]. The quantified Skolemization of our previous example is $\exists f_1 \exists f_2 \forall x \forall z R(x, f_1(x), z, f_2(x, z))$.

Theorem 2.5. *Let φ be a sentence. Then φ is true in exactly the same models as $qsk(\varphi)$.*

This whole Skolemization business depends on something called the *Axiom of Choice*. Since we just use Skolemization to illustrate quantifier interaction (and later on, to introduce non-linear prefixes) we will happily assume the axiom and not worry more about it[4].

Note: it is well known that natural language is very ambiguous. One of the ways in which this ambiguity is present is in quantifier scope, as the following example due to Westerståhl shows: let the sentence "A financial consultant was hired by every company" be called *Sentence 1* and be followed by "He got rich but companies went bankrupt". Then Sentence 1 must be understood with the existential quantifier in front of the universal (we are talking about one single consultant for all the companies). But if Sentence 1 is followed by "Only one of the consultants was still employed one year later" then we must understand it with the universal quantifier in front of the existential quantifier (i.e. one consultant per company, given by a Skolem function from companies to consultants). Note that Sentence 1 could be followed by either sentence equally well.

2.3.3 Quantifier Rank

The *quantifier rank* (sometimes called *quantifier depth*) of a sentence is the number of quantifiers nested inside each other in the sentence. That is, if one represents the sentence as a tree, and counts the number of quantifiers in any path between the root and a leaf, quantifier depth is the maximum such number. Also, if the sentence if put in prenex normal form, quantifier depth is simply the number of quantifiers in the formula.

The formal definition is as follows: $qr(\varphi)$ is defined as

- $qr(\varphi) = 0$ if φ is an atomic (quantifier-free) formula.
- $qr(\varphi_1 \wedge \varphi_2) = qr(\varphi_1 \vee \varphi_2) = max(qr(\varphi_1), qr(\varphi_2))$.
- $qr(\neg\varphi) = qr(\varphi)$.
- $qr(\exists x\varphi) = qr(\forall x\varphi) = qr(\varphi) + 1$.

The reason this notion is important is the following: a variable is really *a name plus a quantifier*. This means that a name alone is not really enough to tell a variable from another. Variables can be *reused*, and a variable can have several *occurrences*. As a result, the number of quantifiers does not equal

---

[3] Note that this is a second-order sentence, since it quantifies over functions.
[4] Further, one can develop Skolemization for *FOL* sentences without the Axiom of Choice, but we will not develop this here (see [66])

the number of *different* variable *occurrences* in a formula (or a sentence!). For instance, the formula

$$\forall x (R(x) \vee S(x))$$

has a single variable (and a single quantifier) that occurs twice. However, the number of quantifiers in a sentence equals the number of distinct variables. In fact, it is the quantifier scope that defines the sentence: if the same variable appears quantified twice, we must consider that there are two variables that happen to have the same name in the sentence. Of course, in order to avoid confusion each variable occurrence can only be within the scope of a single quantifier. An example of variable reuse will make the situation clear: assume, as before, a model with just a binary relation G (that is, a graph). Then, a straightforward way of saying "there is a path of length n between nodes a and b" uses $n-1$ variables (and $n-1$ quantifiers), for a total of $n+1$ variables; for instance, for $n = 5$ we can write

$$\exists x_1 \exists x_2 \exists x_3 \exists x_4 (G(a, x_1) \wedge G(x_1, x_2) \wedge G(x_2, x_3) \wedge G(x_3, x_4) \wedge G(x_4, b))$$

However, we can reuse variables and write the same sentence with only 1 variable:

$$\exists z (G(a, z) \wedge \exists x (G(z, x) \wedge \exists z (G(x, z) \wedge \exists x (G(z, x) \wedge G(x, b)))))$$

In this sentence, the variable x occurs first within the scope of the second quantifier (it has two occurrences in that scope) and a third time under the scope of the fourth quantifier (it also has two occurrences under that scope). Note that when the fourth quantifier occurs, the xs after that belong to it, and the previous variable x is *shadowed* by this new quantifier. This is similar to a variable in a programming language with block scopes being declared twice. Finally, note that both this sentence and the previous one have 4 quantifiers. As we stated before, it is the quantifier scope that causes the variable to be a new one -even under an old name!

For any finite relational vocabulary L, and k, there are, up to logical equivalence, only finitely many first order L-formulas of quantifier rank $\leq k$. This is easy to prove by induction on k.

Variable reuse is important for logics where the number of (distinct) variables is limited. Such logics (called *bounded variable logics* ([81])) are important in the study of computation, since logical variables can be seen in this context as the counterpart of programming variables, which are memory locations and hence resources. Thus, the number of variable *names* tells us about the minimum amount of memory that it takes to compute something (computation in this context means interpreting a formula or sentence in a given model. The games introduced later can be seen as a form of computation). This is why such logics are important in the study of descriptive complexity.

For any k, we define FOL^k the fragment of FOL made up of formulas using at most k variables (either free or bound). Number of variables and quantifier rank will be related on section 2.5.

2.3.4 Relativization

Quantifiers range over the whole domain of a structure. However, many times we want to restrict our attention to some part of the domain. This can be accomplished quite easily in FOL.

Definition 2.6. *Let* $V = (R_1, \ldots, R_n)$ *be a relational vocabulary. A V-structure $\mathcal{A}$ is called a* substructure *of another V-structure $\mathcal{B}$ (in symbols, $\mathcal{A} \le \mathcal{B}$) iff*

- $A \subseteq B$;
- *For all* $i = 1, \ldots, n$, $R_i^{\mathcal{A}} = R_i^{\mathcal{B}} \cap A^{\alpha(R_i)}$.

In vocabularies with constants, it is also demanded that all constants in the vocabulary are present in the substructure.

Lemma 2.7. *Let* $\mathcal{A} \le \mathcal{B}$ *and* ψ *a FOL sentence. If* ψ *is existential, then* $\mathcal{A} \models \psi$ *implies* $\mathcal{B} \models \psi$. *If* ψ *is universal, then* $\mathcal{B} \models \psi$ *implies* $\mathcal{A} \models \psi$.

Given a unary relation (which we called a predicate) $\psi(x)$, we say that, given structure $\mathcal{A}$ and interpretation s, the set $\{x \mid \psi(x)\}$ denotes the set of elements $a \in A$ such that $\mathcal{A} \models \psi[s(x/a)]$.Note that $\{x \mid \psi(x)\} \subseteq A$. Note also that, since this set works like the existential quantifier, and we assume that there are no other *free* variables in ψ, the definition is really independent of the interpretation s, and it depends only on ψ and $\mathcal{A}$.

Assume fixed a structure $\mathcal{A}$. The relativization of a FOL sentence φ to some set $B \subseteq A$ is accomplished by: one, finding a formula with one free variable $\psi(x)$ such that $\{x \mid \psi(x)\} = B$; and two, changing φ as follows:

- any sub-formula of φ of the type $\exists x\ \rho(x)$ is changed to $\exists x\ (\psi(x) \wedge \rho(x))$;
- any sub-formula of φ of the type $\forall x\ \rho(x)$ is changed to $\forall x\ (\psi(x) \rightarrow \rho(x))$.
- any other sub-formula of $|varphi$ is left unchanged.

The new formula, call it φ^B, states the same thing about B that the original formula φ stated about the whole domain, A. This idea is formalized in the next theorem. First, note that given structure $\mathcal{A}$ and set $B \subseteq A$, we can generate a substructure $\mathcal{B}$ of $\mathcal{A}$ with domain exactly B.

Theorem 2.8. *Let* $\mathcal{A}$ *be a structure,* $\mathcal{B} \le \mathcal{A}$ *with domain B. Then, for all FOL sentences φ,*

$$\mathcal{A} \models \varphi \text{ iff } \mathcal{B} \models \varphi^B$$

In fact, the most frequent use of quantification is in relativized form, since in complex domains we rarely make sweeping statements about the whole domain. Take, for instance, the structure $\mathcal{N} = <N, +, *, 0, 1>$, where N is the set of natural number, $+$ and $*$ are the standard addition and multiplication operations, and 0 and 1 are the constants denoting zero and one. Then, some general statements are

$$\forall x \ (x + 0 = x)$$

$$\forall x \ (x * 1 = x)$$

Many statements of interest in number theory, though, are relativized. Let the predicate $D(x, y) = \exists z(x * z = y)$, which obviously means that x divides y; and let the predicate $P(x) = \neg \exists z \ (z \neq 1 \land z \neq x \land D(z, x))$, which means that x is prime. If we want to make some statement about primes, we write a formula of the type

$$\forall x(P(x) \rightarrow \varphi(x))$$

Additional restrictions are, of course, possible. For instance, Euclid's lemma can be written as

$$\forall x \forall y(P(x) \land D(x, y) \land \exists z \ \exists w \ (y = z * w)) \rightarrow D(x, z) \lor D(x, w)$$

2.4 Games

There is another, very intuitive way to the semantics of sentences. It is based on *two-person games*, due to Ehrenfeucht and Fraisse (and sometimes called *Ehrenfeucht-Fraisse games*).

Consider a sentence like $\forall x \exists y \varphi(x, y)$. The sentence is either true or false in a given structure $\mathcal{A}$. We can see establishing its truth value as a game between two players, Spoiler (Abelard) and Duplicator (Eloise), each one taking on side: Spoiler tries to prove the sentence false and Duplicator tries to prove the sentence true. The moves on the game consist of picking values for the variables in the sentence from among the elements in the domain of $\mathcal{A}$. Because of quantifier semantics, Spoiler picks values for universally quantified variables (trying to find one that falsifies the sentence), while Duplicator picks values for existentially quantified variables (trying to find the right one that makes the sentence true. Such value is usually called a **witness**).

The game to determine the semantics of a *FOL* sentence φ (i.e. whether it is true or false in structure $\mathcal{A}$) is played as follows:

- if $\varphi = \neg \psi$, Spoiler and Duplicator exchange roles.
- if $\varphi = \psi_1 \lor \psi_2$, the Duplicator chooses one of ψ_1, ψ_2 to continue the game.
- if $\varphi = \psi_1 \land \psi_2$, the Spoiler chooses one of ψ_1, ψ_2 to continue the game.
- if $\varphi = \exists x \psi(x)$, the Duplicator chooses an individual $a \in A$ and the game continues with $\psi[a/x]$.
- if $\varphi = \forall x \psi(x)$, the Spoiler chooses an individual $a \in A$ and the game continues with $\psi[a/x]$.
- if φ is atomic, the game ends. The Duplicator wins if φ is true in $\mathcal{A}$; otherwise, the Spoiler wins. Note that by the time we get to an atomic formula, all free variables should have been substituted by elements of the structure -otherwise φ was not a sentence to start with. Hence we can determine truth value easily.

A player has a winning strategy if there is a rule that tells that player how to move in such a way that, no matter what the other player does, this player wins the game.

Note that the game reflects the quantifier scope and dependence, as follows: if an existential quantifier precedes a universal one, Duplicator chooses an individual without knowing what Spoiler will choose in turn. Thus, the individual must be good for all individuals, as Spoiler may choose any element of the domain. If a universal quantifier precedes a universal one, Spoiler chooses an individual first, and the Duplicator can make a choice *based on what Spoiler did*, that is, Duplicator's choice depends on Spoiler. It is this dependence that Skolem functions capture formally.

Games are usually played in two structures, A and B, of a common vocabulary L. Let r be a positive integer. The game of length r associated with A and B is played as follows: Spoiler starts and picks an element from either A or B; Duplicator follows by picking an element in the other structure (the one Spoiler didn't chose from). This is a move, and is repeated r times. On each move, Spoiler choses both the structure and the element, while Duplicator only choses the element. Let a_i the element in A picked in the ith move, and b_i the element in B picked in the ith move. The set of pairs $\{(a_1, b_1), \ldots, (a_r, b_r)\}$ is a *round* of the game. Duplicator wins the round if the mapping $a_i \to b_i$ is a *partial isomorphism* between A and B, that is, an isomorphism between the substructure of A generated by $\{a_1, \ldots, a_r\}$ and the substructure in B generated by $\{b_1, \ldots, b_r\}$. Duplicator wins the game of length r if it has a winning strategy. Games simulate back-and-forth equivalence of depth r.

The main interest of the game is the following: if Duplicator has a winning strategy for games of length r, then one cannot distinguish between structures A and B by looking at r elements at a time. In particular, the structures are not distinguishable by sentences of quantifier depth $\leq r$.

Basically, number of pebbles corresponds to number of variables, and number of moves corresponds to nesting depth of quantifiers. This will be formalized in the next section.

2.5 More Semantics

Here we describe a few more semantic concepts that are tied to quantification. First, we introduce some basic relations among structures.

Definition 2.9. *Let* $\mathcal{A} =< A, R_1, \ldots, R_m >$ *and* $\mathcal{B} =< B, R'_1, \ldots, R'_m >$ *be two L-structures. A* homomorphism *between* $\mathcal{A}$ *and* $\mathcal{B}$ *is a function* $h : A \to B$ *such that for all* $a_1, \ldots, a_k \in dom(h)$, *if* $R_i(a_1, \ldots, a_k)$ *then* $R'_i(h(a_1), \ldots, h(a_k))$.

Definition 2.10. *A homomorphism* $h : A \to B$ *is an* isomorphism *if its inverse is also a homomorphism. Note that this implies that an isomorphism is a bijection between A and B. If there exists an isomorphism between* $dom(\mathcal{A})$ *and* $dom(\mathcal{B})$, *we say that* $\mathcal{A}$ *and* $\mathcal{B}$ *are* isomorphic *(in symbols,* $\mathcal{A} \cong \mathcal{B}$).

Definition 2.11. *Let $\mathcal{A}$ and $\mathcal{B}$ to L-structures that satisfy exactly the same FOL sentences (that is, for each FOL sentence ψ, $\mathcal{A} \models \psi$ iff $\mathcal{B} \models \psi$). Then $\mathcal{A}$ and $\mathcal{B}$ are called* elementarily equivalent, *written $\mathcal{A} \equiv \mathcal{B}$.*

Definition 2.12. *For any two structures $\mathcal{A}$ and $\mathcal{B}$,*

- $\mathcal{A} \equiv^k \mathcal{B}$ iff $\mathcal{A}$ and $\mathcal{B}$ agree on all formulas in FOL^k.
- $\mathcal{A} \equiv_p \mathcal{B}$ iff $\mathcal{A}$ and $\mathcal{B}$ agree on all FOL sentences of quantifier rank $\leq p$.
- $\mathcal{A} \equiv_p^k \mathcal{B}$ iff $\mathcal{A} \equiv^k \mathcal{B}$ and $\mathcal{A} \equiv_p \mathcal{B}$.

All the above definitions are related; here we give only the most basic (and important) relations.

Lemma 2.13. *(Homomorphism theorem) Let f be a homomorphism between structures $\mathcal{A}$ and $\mathcal{B}$, $\varphi(x_1, \ldots, x_n)$ a FOL formula and $a_1, \ldots, a_n \in A$. Then $\mathcal{A} \models \varphi[a_1, \ldots, a_n]$ iff $\mathcal{B} \models \varphi[f(a_1), \ldots, f(a_n)]$. In particular, for any FOL sentence ψ (that is, when $n = 0$), $\mathcal{A} \models \psi$ iff $\mathcal{B} \models \psi$.*

Corollary 2.14. *Isomorphic structures are elementarily equivalent. That is, if $\mathcal{A} \cong \mathcal{B}$), then $\mathcal{A} \equiv \mathcal{B}$.*

Lemma 2.15. *For any k, if $\mathcal{A} \equiv^k \mathcal{B}$ then $\mathcal{A} \equiv_k \mathcal{B}$.*

Lemma 2.16. *Let $\mathcal{A}, \mathcal{B}$ be two structures. Then Duplicator has a winning strategy in the p-round E-F game with k pebbles iff $\mathcal{A} \equiv_p^k \mathcal{B}$*

Lemma 2.17. *Let $\mathcal{A}, \mathcal{B}$ be two structures. Then Duplicator has a winning strategy in the p-round E-F game iff $\mathcal{A} \equiv_p \mathcal{B}$*

Lemma 2.18. *Let $\mathcal{A}, \mathcal{B}$ be two structures. Then Duplicator has a strategy for winning indefinitely in E-F game with k pebbles iff $\mathcal{A} \equiv^k \mathcal{B}$*

This last result can be reinterpreted with structures. To do so, we note that isomorphisms are functions relating whole structures, but we can give a more local definition, which is interesting on its own right.

Definition 2.19. *Let $\mathcal{A} = <A, R_1, \ldots, R_m>$ and $\mathcal{B} = <B, R'_1, \ldots, R'_m>$ be two L-structures. A partial (local) isomorphism of size n between $\mathcal{A}$ and $\mathcal{B}$ is a bijective function h with $dom(h) \subseteq A$, $rng(h) \subseteq B$, $|dom(h)| = n$, and such that for all $a_1, \ldots, a_k \in dom(h)$, $R_i(a_1, \ldots, a_k)$ iff $R'_i(h(a_1), \ldots, h(a_k))$. Note: the empty map is a partial isomorphism.*

This should be easy to see:

Lemma 2.20. *Let $\mathcal{A} = <A, R_1, \ldots, R_m>$ and $\mathcal{B} = <B, R'_1, \ldots, R'_m>$ be two L-structures, and f a partial isomorphism of size n (in which case we say that $\mathcal{A}$ and $\mathcal{B}$ are n-isomorphic). Then for every formula φ of FOL with n or less free variables,*
$$\mathcal{A} \models \varphi(a_1, \ldots, a_n) \ iff \ \mathcal{B} \models \varphi(f(a_1), \ldots, f(a_n))$$

Partial isomorphisms preserve truth for atomic sentences (quantifier free). However, they do not preserve truth for formulas with quantifiers. The intuitive reason is that atomic sentences talk about particular elements, while sentences with quantifiers talk about arbitrary elements. For example, assume our language is a binary relation symbol R. Then the atomic formula $R(a, b)$, where a, b are constants, will be true in a structure $\mathcal{A}$ iff it is true in a structure (of similar type) $\mathcal{B}$, whenever $\mathcal{A}$ and $\mathcal{B}$ are related by a partial isomorphism that sends a and b to some elements in the domain of $\mathcal{B}$ (otherwise the function would not be a partial isomorphism). However, formula $\exists x R(a, x)$ only states that element a is R-related to some other element x; a partial isomorphism may map a and some other domain element b to elements in $\mathcal{B}$, but there is on guarantee that the other element is R-related to a (if it is not in $\mathcal{A}$, then it will not be in $\mathcal{B}$).

A partial isomorphism can always be extended to a (total) isomorphism if the structures are isomorphic.

A *back-and-forth* system is a set of partial isomorphism of size n for each n. That is, we can build partial isomorphisms of size n no matter which elements we choose.

Definition 2.21. *Let $\{I_n\}_{n=0}^{\infty}$ be a family of non empty sets of partial isomorphisms from $\mathcal{A}$ to $\mathcal{B}$. Then $\{I_n\}_{n=0}^{\infty}$ has the* Back and Forth *property if*

- *(Simple Forth) For every $f \in I_n$ and every $a \in A$, there is a $g \in I_{n-1}$ extending f such that $a \in Dom(g)$.*
- *(Simple Back) For every $f \in I_n$, and every $b \in B$ there is a $g \in I_{n-1}$ extending f such that $b \in Rng(g)$.*

We say that $\mathcal{A}$ and $\mathcal{B}$ are finitely isomorphic iff there is a family $\{I_n\}_{n=0}^{\infty}$ of non empty sets of partial isomorphisms from $\mathcal{A}$ to $\mathcal{B}$ that has the Simple Back and Forth property. If there is only a finite family $\{I_n\}_{n=0}^{m}$ of non empty sets of partial isomorphisms from $\mathcal{A}$ to $\mathcal{B}$ that has the Simple Back and Forth property, we say that $\mathcal{A}$ and $\mathcal{B}$ are m-equivalent.

Note that this is similar to the definition of E-F games, and hence to equivalence up to formulas of quantifier rank m.

Definition 2.22. *Two structures $\mathcal{A}$ and $\mathcal{B}$ are m-isomorphic iff $\mathcal{A} \equiv_m \mathcal{B}$.*

Theorem 2.23. *$\mathcal{A}$ is m-isomorphic (for the Simple Back and Forth property) to $\mathcal{B}$ iff $\mathcal{A}$ is m-equivalent to $\mathcal{B}$.*

Theorem 2.24. *(Fraisse's Theorem) Given a finite language L, two L-structures are elementarily equivalent iff they are finitely isomorphic.*

Fraisse's Theorem is proven by noticing that any FOL sentence has a (finite) m rank, and looking at m-equivalences. With a finite number of symbols in the language, there are only finitely many formulas of quantifier rank m (up to isomorphism), and therefore one can build appropriate m equivalences.

2.5.1 Expressive Power of *FOL*

The games above are used to prove that some property is not expressible in *FOL*, as follows: if it were, there would be a *FOL* sentence φ that would express it. This sentence has a finite quantifier depth, r. Then pick two structures A and B such that A has the property, B does not, but you need to look at more than r elements to determine that. Then you can give a winning strategy for Duplicator for games of length r.

Using the above, one can prove that many properties (connectivity of a graph, evenness, 2-colorability of graphs, Eulerian graphs[5]) are not expressible in *FOL*.

As an example, let the structure be a graph, and a, b two nodes. There is no sentence that will say that a has as many neighbors as b (same number). We can, for any n, take two graphs, one where a has n+2 neighbors and b has n+1 neighbors. Because there are only n rounds, we can never show the difference.

As another example, think of graph connectedness. For each n, take two graphs, one being a cycle of length 2^n and another one made of two disjoint cycles of length 2^n. At every round j (for all $j \le n$), the Duplicator chooses nodes such that the distance with any previous node in that structure is at least 2^{n-j}. This is always possible, due to the sizes of the graphs. Spoiler cannot show that there are two nodes with no path between them in the second structure, because all paths can have at most length n.

A final example is transitive closure. If (A, R) is a graph, and we could define transitive closure with formula $\psi(x, y)$, then we could define connectedness with the sentence $\forall x \; \forall y \; \psi(x, y)$.

2.5.2 Finite and Infinite Models

Usually it is considered that structures have infinite domains. When one studies models, it is very convenient to do so. However, in the context of databases it is also very interesting to think about the finite case. It turns out that finite and infinite models are *very* different, for reasons we cannot explain here. The study of finite models is a fascinating subject that has taken off in the last few years, but here we can only point out a few things that are of interest to us later.

Let $\mathcal{A}$ be a *finite* model, that is, a model with a finite domain A (note that this implies that all relations and functions in $\mathcal{A}$ are also finite). Let the cardinality of A be n. Then, for a universal sentence like
$$\forall x \varphi(x)$$
we can come up with a sentence that is equivalent $\mathcal{A}$:
$$\bigwedge_{a \in A} \varphi(a)$$

---

[5] A graph is Eulerian if there exists an Euler cycle on it, i.e. a cycle passing through each edge exactly once.

That is, if the elements of A are called $a_1, a_2, \ldots$, this is the conjunction
$\varphi(a_1) \wedge \varphi(a_2) \wedge \ldots = \bigwedge_{i=1}^{n} \varphi(a_i)$

The important thing to note is that this is a *finite* conjunction, and therefore it is a *FOL* formula. Some logic admit infinite conjunctions and disjunctions and formulas (not surprisingly, they are called *infinitary logics*). Likewise, for an existential sentence like

$\exists x \varphi(x)$

we can come up with a sentence that is equivalent $\mathcal{A}$:

$\bigvee_{a \in A} \varphi(a)$

That is, if the elements of A are called $a_1, a_2, \ldots$, this is the disjunction
$\varphi(a_1) \vee \varphi(a_2) \vee \ldots = \bigvee_{i=1}^{n} \varphi(a_i)$

Repeating the process recursively, we can get rid of all the quantifiers in the original sentence.

Note, though, that these sentences are equivalent to the original ones with quantifiers only *in the given model* $\mathcal{A}$. For a different finite model $\mathcal{B}$, we need a different sentence. We can see now that the power of quantification is that it let us say something about all or some of the elements in a model *regardless of the model.* As an example of this power, one may wonder if there is a smarter way to get rid of quantifiers that works in all cases. The answer is no, there is no way to always get rid of the quantifiers -although there are procedures for *quantifier elimination* in some cases, and it is very interesting to know in which cases the elimination is possible and in which cases it is not. The point we want to make here, though, is that quantifiers truly contribute to the expressive power, and that is why they cannot, in general, be eliminated.

However, the fact that any *FOL* sentence has a finite m rank means that we can only "check" m elements of a structure to determine truth or falsity of the sentence; it turns out that this is already quite strong in finite models:

Theorem 2.25. *Finitely isomorphic finite models are isomorphic.*

Intuitively, if any (arbitrary) m elements in one model can be put in a isomorphic relation with m elements in the other models, then the relation can be extended to *all* the elements.

We saw earlier that (clearly) if $\mathcal{A} \cong \mathcal{B}$, then $\mathcal{A} \equiv \mathcal{B}$. In finite models, the converse is true, so we have this:

Theorem 2.26. *For finite models $\mathcal{A}$ and $\mathcal{B}$, $\mathcal{A} \equiv \mathcal{B}$ iff $\mathcal{A} \cong \mathcal{B}$.*

In the rest of the book, we will consider finite models, since these are the most appropriate models for considering computational issues, and our focus will be in efficient implementation of quantification, which means that we will try to give effective procedures for computing sentences with quantifiers. Working with finite models will provide a definitive advantage here!

3

Generalized Quantifiers

3.1 Introduction

Generalized Quantifiers (henceforth GQs) were originally introduced by Mostowski ([76]). Lindstrom generalized the original idea and set the definition used nowadays ([70])[1]. GQs remained mostly a concept of interest for logicians until years later. In a seminal paper, Barwise and Cooper showed their importance in natural language formalization ([16]). Following in a tradition started by Montague, Barwise and Cooper show how GQs are a useful and powerful tool in the formal analysis of linguistic phenomena. This insight established an interesting area of research, which we discuss briefly in chapter 8. Also, work on *descriptive complexity* has produced recently a large body of results tied to GQs (see [47, 65, 93] as a small sample). Finally, some work has focused on practical applications of the concept in diverse areas of Computer Science, like query languages ([52, 85]), Artificial Intelligence ([90], [82]) and graphical interfaces ([17]). The research described in this book continues in this tradition, focusing on practical uses in query languages, and Information Systems in general.

In this chapter we introduce the concept of GQ formally, provide several examples and describe some basic properties. Here we focus on fixing notation and providing some basic facts that will be used in the rest of the book. Additional properties will be introduced later as needed.

3.2 Generalized Quantifiers

There are two views on the concept of GQ, as relations and as classes of structures. We introduce first the former, as it may be a bit more intuitive, and provide some examples. Then we develop the latter definition in order to introduce some notation that will be of use later. Of course, both definitions

[1] *GQs* are also called *Lindstrom quantifiers.*

A.Badia, *Quantifiers in Action: Generalized Quantification in Query, Logical and Natural Languages*, Advances in Database Systems 37, DOI: 10.1007/978-0-387-09564-6_3,
© Springer Science + Business Media, LLC 2009

are equivalent, and we will use one or the other throughout the book as a matter of convenience.

We consider fixed an infinite *universe* $\mathcal{U}$, the set of all values. We use M, M' to indicate *domains*, that is, non-empty subsets of $\mathcal{U}$, and $A, A_1, \ldots, B, B_1, \ldots$ to indicate subsets of M^n, for a given domain M and positive natural number n.

Definition 3.2.1 *Let a* type *be a finite sequence of positive numbers, written* $(k_1, \ldots, k_n)$.

- *A global generalized quantifier Q of type $(k_1, \ldots, k_n)$ is a function that assigns, to any set (domain) $M \subseteq \mathcal{U}$, a local generalized quantifier Q of type $(k_1, \ldots, k_n)$ on M.*
- *A local GQ of type $(k_1, \ldots, k_n)$ on M is an n-ary relation between subsets of $M^{k_1}, \ldots, M^{k_n}$ (i.e. between elements of $\mathcal{P}(M^{k_1}) \times \ldots \times \mathcal{P}(M^{k_n}))$*

Further, Q is closed under bijections: if $(A_1, \ldots, A_n)$ with $A_i \subseteq M^{k_i}$, $1 \leq i \leq n$ are in the extension of Q in M, and $f : M \to M'$ is a bijection, then $(f[A_1], \ldots, f[A_n])$ with $f[A_j] \subseteq M'^{k_j}$, $1 \leq j \leq n$ and $M' \subseteq \mathcal{U}$, is in the extension of Q in M'.

We will use Q as a variable over GQs, and boldface for names of particular quantifiers. We will write Q_M^t to indicate the quantifier of type t in domain M, although later on we may drop the subscript or the superscript as we will fix one or the other for some chapters. The expression $Q_M^t(A_1, \ldots, A_n)$ indicates that sets $A_1, \ldots, A_n$ belong to the extension of Q in universe M, i.e. that they are in the relation denoted by Q.

Next we introduce several examples of GQs, including some common ones that will be used later in queries and lemmas. To simplify notation, we consider a domain M as fixed and give the local definition of the quantifier, which allows us to drop the subscript. We keep the superscript as some quantifiers share the same name while having different types; these can be considered as versions of the same idea, and they are related in a way that will be explained later on.

$$\mathbf{all}^{(1,1)} = \{A, B \subseteq M | A \subseteq B\}$$

$$\mathbf{all}^{(1)} = \{A \subseteq M | A = M\}$$

$$\mathbf{C}^{(1)} = \{A \subseteq M | |A| = |M|\} \text{ Chang's quantifier}$$

$$\mathbf{some}^{(1,1)} = \{A, B \subseteq M | A \cap B \neq \emptyset\}$$

$$\mathbf{some}^{(1)} = \{A \subseteq M | A \neq \emptyset\}$$

$$\mathbf{no}^{(1,1)} = \{A, B \subseteq M | A \cap B = \emptyset\}$$

$$\mathbf{no}^{(1,1)} = \{A \subseteq M | A = \emptyset\}$$

$$\mathbf{not\ all}^{(1,1)} = \{A, B \subseteq M | A \nsubseteq B\}$$

$$\mathbf{at\ least}_n^{(1,1)} = \{A, B \subseteq M | |A \cap B| \geq n\}$$

$$\mathbf{at\ least}_n^{(1)} = \{A \subseteq M | |A| \geq n\}$$

$$\mathbf{at\ most}_n^{(1,1)} = \{A, B \subseteq M | |A \cap B| \leq n\}$$

$$\mathbf{at\ most}_n^{(1)} = \{A \subseteq M | |A| \leq n\}$$

$$\mathbf{more}^{(1,1)} = \{A, B \subseteq M | |A| > |B|\}^2$$

$$\mathbf{most}^{(1,1)} = \{A, B \subseteq M | |A \cap B| > |A - B|\}$$

$$\mathbf{most}^{(1)} = \{A \subseteq M | |A| > |M - A|\}$$

$$\mathbf{I}^{(1,1)} = \{A, B \subseteq M | \ |A| = |B|\} \text{ (Hartig's quantifier)}$$

$$\mathbf{Q}_\omega^{(1)} = \{A \subseteq M | \ |A| = \omega\} \text{ (Mostowski's quantifier)}$$

$$\mathbf{P_{m,n}} = \{X, Y \subseteq M \ | \ |X \cap Y| > |X| \times \frac{m}{n}\}$$

$$\mathbf{Q}_{Eul}^{(2)} = \{E \subseteq M^2 \ | \ \forall x \ |\{y \ | \ E(x,Y)\}| \text{is even})\}$$

$$\mathbf{H}^{(4)} = \{R \subseteq M^4 | \exists f : M \to M \ \exists g : M \to M \ \forall a, b \in M < a, f(a), b, g(b) > \in R\}$$

$$\mathbf{W^{R}}^{(1,2)} = \{A \subseteq M, \ R \subseteq M^2 \ | \ R \text{ well-orders } A\}$$

$$\mathbf{Q_E} = \{A \subseteq M \ | \ |A| \text{ is even}\}$$

Note that $\mathbf{all}^{(1)}(A)$ implies $\mathbf{C}(A)$, but not the other way around -the counterexample being infinite sets. Quantifiers of type (2) can be identified with *graph properties*, in the sense that they give properties of binary relations and we can identify binary relations with graphs[3]. Thus, $\mathbf{Q}_{Eu}$ denotes Eulerian graphs (using the well known property that a connected graph is Eulerian iff every vertex has an even degree). We could also define other graph properties, e.g. $Q_{Ham}^{(2)}$ which holds true only of Hamiltonian graphs. Some authors use the name *Rescher's* quantifier for $\mathbf{more}^{(1,1)}$ (see [93]), while others offer the definition given here. $\mathbf{Q}_\omega$ is the original quantifier defined by Mostowski, and is meant to denote infinite sets (here ω is the first infinite cardinal). $\mathbf{I}$ is a simple example of a property that is not first order-definable, while the quantifier $\mathbf{H}$ is a more complex (and powerful) one[4]. Quantifier $\mathbf{most}^{(1)}$ can

---

[3] Note that most of the times a graph is assumed to be loop-free and non-directed, and hence the corresponding binary relation must be irreflexive and symmetric.

[4] $\mathbf{H}$ be used in chapter 6 to capture *non linear prefixes*.

be paraphrased as **more than half of everything**. The quantifiers of type $\mathbf{P}_{m,n}$ are sometimes called the *proportional* quantifiers. Note that the definition of **aleast**$_n$, **at most**$_n$ and the proportional quantifiers are definitions with parameters, and hence each one denotes a set of quantifiers (when we instantiate n -and m if needed- to two natural numbers, we obtain a particular quantifier). $\mathbf{Q_E}$ is an interesting quantifier in that it is very simple, can be added to FOL in a straightforward way, and yet it transforms the logic radically, since it is known that FOL has a *0-1 law* on finite models, while the logic with $\mathbf{Q_E}$ added obviously does not have this law[5].

The length of a type is called the *binding* (of quantifiers of that type). We will talk about $n - place$ or $n - ary$ quantifiers to refer to the *binding*. The greatest element in the type is the *arity* of the quantifier. We will talk about $m - adic$ quantifiers to refer to the *arity* of the quantifier. Thus, GQs of type $(1, \ldots, 1)$ are *monadic GQs* and they can be unary (type (1)), binary (type $(1,1)$), etc. In this text, unary monadic quantifiers (type $[1]$) are called *simple*, and binary monadic quantifiers (type $(1,1)$) are called *standard*. When the arity is > 1 we talk of *polyadic* quantifiers. All the above quantifiers (except $\mathbf{Q}_\omega$, $\mathbf{Q_R}$ and $\mathbf{H}$, which are unary) are binary. They are all *monadic* except $\mathbf{W^R}$, which has type $[1, 2]$, and $\mathbf{H}$, which has type $[4]$.

Note that not every relation between subsets of the domain is considered a GQ. Intuitively, we would like a GQ to behave as a *logical* operator, in the sense that it should not distinguish between elements in the domain, and nothing should be dependent on the domain M chosen. Hence our requirement that GQs are closed under bijections. In the context of database query languages, this constraint ensures that quantifiers are *generic* operations, while EXT ensures its domain independence. Both are important properties for query operators ([6]). We will exploit them in chapter 4 to analyze issues of safety and domain independence.

Some authors express express this requirement as a separate axiom, and give it a name ([102]):

Definition 3.2.2 (ISOM) A quantifier Q follows ISOM if, whenever f is a bijection from M to M', then $Q_M(A_1, \ldots, A_n)$ iff $Q_{M'}(f[A_1], \ldots, f[A_n])$.

In the context of monadic quantifiers, ISOM implies that only the sizes of the sets, as opposed to the concrete individuals, matter. That is why it is called QUANT in [97], where only one domain is considered and therefore

---

[5] A property has a 0-1 law on finite models if, as the size n of the model grows, the probability that the property holds in all models of size n goes to 0 or to 1. A logic has a 0-1 law on finite models if all the properties definable in the logic have a 0-1 law. This is a strong characterization of FOL, but it fails with $\mathbf{Q_E}$ since clearly the property of being even *flip-flops* between 0 and 1 as n grows, therefore not converging to either. [25] studies 0-1 laws for extensions of first-order logic with generalized quantifiers (although they restrict themselves to type (2) since they study graph properties.)

permutations take the place of bijections (the same happens in [91]). That is the reason that this axiom is also called $PERM$:

Definition 3.2.3 (PERM) A quantifier Q follows PERM if for any domain M, and any f permutation on M, $Q_M(A_1, \ldots, A_l)$ iff $Q_M(f(A_1), \ldots, f(A_l))$.

There have been several other *axioms* considered. Barwise and Cooper ([16]) propose, for natural language quantifiers, the following one:

Definition 3.2.4 (CONSERV) A quantifier Q of type $[k, k_1, \ldots, k_n]$ follows CONSERV if for all M and all $A \subseteq M^{k_1}, A_1 \subseteq M^{k_1}, \ldots, A_n \subseteq M^{k_n}$, $Q_M(A, A_1, \ldots, A_n)$ iff $Q(A, A_1 \cap A, \ldots, A_n \cap A)$.

They call this *living on a set*. Other authors call quantifiers that obey this axiom *conservatives*. The axiom gives the first argument a privileged role. Natural language analysis has convinced researchers that this is indeed a very useful axiom, and it is assumed in other studies of generalized quantifiers in linguistics ([96, 16, 61]). However, from a purely formal perspective it creates an *imbalance* which is not justified in logical grounds. For instance, quantifiers like **I** and **more** are *not* conservative.

CONSERV implies that what happens outside the sets $A, A_1, \ldots, A_n$ is irrelevant. This *partially* implies that the behavior of a quantifier is independent of the context, as is the case for the usual logic constants. This seems, therefore, a desirable property. There is another axiom that, without giving priority to any argument, makes context independence a characteristic:

Definition 3.2.5 (EXT) A quantifier Q follows EXT if for all $M, M, 'A, ._1. ., A_n \subseteq M \subseteq M'$, $Q_M(A_1, \ldots, A_n)$ iff $Q_{M'}(A_1, \ldots, A_n)$.

While CONSERV almost restricts the domain of quantification to the union of the first n arguments, the determiner may still depend on the domain M. EXT expresses that quantifiers are completely independent of the domain, as the following proves:

Proposition 3.2.6 ([102]) Under EXT, ISOM and PERM are equivalent.

The idea is that under EXT the domain is irrelevant, and therefore thinking of bijections between two domains or permutations inside one capture the same notion. We will say that quantifiers that follow EXT are *domain independent*.

There are yet other axioms that have been proposed; for instance, in linguistics axioms aim to capture *semantic universals* (i.e. rules that seem to be followed by all natural language determiners). In choosing which axioms to incorporate into a definition of GQ, we have to keep in mind that it is not clear that there is a unique set of axioms that are better for all aspects of query languages. We know that operators in query languages need to be generic, and because of genericity we will always assume, when talking about query languages, that quantifiers are generic (i.e. follow ISOM). But, in order to be as general as possible, we will state, at the beginning of each section, any additional axioms we assume or need.

We close this section by providing a definition that has been found useful in the literature, and for which we will have some use later:

Definition 3.2.7 (Monotonicity) An n-ary quantifier Q is *downward monotone* on the i-th argument ($1 \leq i \leq n$) if whenever $Q(A_1, \ldots, A_i, \ldots, A_n)$ and $B \subseteq A_i$, it is the case that $Q(A_1, \ldots, A_{i-1}, B, A_{i+1}, \ldots, A_n)$. An n-ary quantifier is downward monotone iff it is downward monotone on the i-th argument for some i ($1 \leq i \leq n$). Similar definitions are used for *upward monotone* on the i-th argument and *upward monotone*.

3.3 Another view

In the following, the concept of *logic* is taken in a general sense; in any case it implies that the language is question is *at least as powerful* as FOL[6].

Our goal here is to show how to incorporate a quantifier Q into a logic $\mathcal{L}$. To facilitate this task, we introduce the following, equivalent definition of generalized quantifier.

Definition 3.3.1 A global generalized quantifier Q of type $(k_1, \ldots, k_n)$ is again a function that assigns, to each non-empty domain M, a local generalized quantifier of that type on M. A local generalized quantifier Q on M is a class of structures $\mathcal{K}$ such that the vocabulary of all structures on $\mathcal{K}$ is made up of n relation symbols $R_1, \ldots, R_n$, each R_i is of arity k_i, M is the domain, $< M, R_1, \ldots, R_n > \in \mathcal{K}$ iff $Q(R_1, \ldots, R_n)$ and $\mathcal{K}$ is closed under isomorphisms.

In this definition we assign, to a given set, a collection of structures. The last conditions is the expression of ISOM in the present context. Sometimes we will look at GQs as relations in a domain and sometimes as classes of structures, depending on what is notationally or technically more convenient in a given context. In the following, the class notation arises naturally.

Given a logic $\mathcal{L}$, and a GQ Q of type $(k_1, \ldots, k_n)$, we can form the logic $\mathcal{L}(Q)$ by *adding* the quantifier to the logic. For this we need to add a syntactic rule and a semantic rule:

- If $\phi_1(x_1^1, \ldots, x_{k_1}^1), \ldots, \phi_n(x_1^n, \ldots, x_{k_n}^n)$ are formulas in $\mathcal{L}$ such that the variables in $x_1^i, \ldots, x_{k_i}^i$ are pairwise distinct for $1 \leq i \leq n$[7], then

$$Q x_1^1, \ldots, x_{k_1}^1, \ldots, x_1^n, \ldots, x_{k_n}^n (\phi_1(x_1^1, \ldots, x_{k_1}^1), \ldots, \phi_n(x_1^n, \ldots, x_{k_n}^n))$$

  is in $\mathcal{L}(Q)$.

---

[6] See [28] and [29] for a formalization of this idea. Technically, we are talking about *regular logics*.

[7] We only display the variables relevant to the definition. That is, $x_1^i, \ldots, x_{k_i}^i$ is a subset of the set of free variables of ϕ_i, for $1 \leq i \leq n$.

- $\mathcal{M} \models Qx_1^1,\ldots,x_{k_1}^1,\ldots,x_1^n,\ldots,x_{k_n}^n(\phi_1(x_1^1,\ldots,x_{k_1}^1),\ldots,\phi_n(x_1^n,\ldots,x_{k_n}^n))$
  iff
  $(\phi_1^M,\ldots,\phi_n^M) \in Q(M)$, where $\phi_i^M = \{a_1,\ldots,a_{k_i} \mid \mathcal{M} \models \phi_i[a_1,\ldots,a_{k_i}]\}$
  $(1 \le i \le n)$[8].

Syntactically, a quantifier of type $(k_1,\ldots,k_n)$ binds n formulas and a total of $k_1 + \ldots + k_n$ variables. It is clear now that the semantic rule indicates that Q gives raise to a class of structures $< M, \phi_1^M, \ldots, \phi_n^M >$, hence our new definition.

Definition 3.3.2 Let $\mathcal{L}$ be any logic. We say that a Generalized Quantifier of type $[k_1,\ldots,k_m]$ is *definable* in $\mathcal{L}$ iff there is a formula ψ with $z = \sum_{i\le m} k_i$ variables and predicate names $P_i,\ldots,P_m$, where P_i has arity k_i $(1 \le i \le m)$, such that for all models $\mathcal{M}$, $\mathcal{M} \models Q(A_1,\ldots,A_m)$ iff $\mathcal{M} \models \psi(P_1,\ldots,P_m,x_1,\ldots,x_z)$.

When adding a quantifier to a language, we are trying to enable the language to express a certain *property*. However, if we see a GQ as representing a certain structural property ([80]), it is natural that this property be represented in any arity, so that the logic has a certain closure with respect to the property.

The first definition that is relevant here is that of *relativization*.

Definition 3.3.3 If Q is of type $(k_1,\ldots,k_n)$, the relativization of Q is the generalized quantifier Q^r of type $(1,k_1,\ldots,k_n)$, defined by (for all $A \subseteq M$ and $R_i \subseteq M^{k_i}$)

$$< A, R_1,\ldots,R_n >\in Q_M^r \Longleftrightarrow < R_1 \cap A^{k_1},\ldots,R_n \cap A^{k_n} >\in Q_A$$

In other words, for $A \subseteq M$ we can use Q^r to express in M what Q says in A. We have already seen several examples of relativization, since for any M, any $A \subseteq M$, **some**$^{(1)}(A)$ iff **some**$^{(1,1)}(M,A)$, and **all**$^{(1)}(A)$ iff **all**$^{(1,1)}(M,A)$. Thus, **all** of type $(1,1)$ is a relativization of $\forall$; while **some** of type $(1,1)$ is a relativization of $\exists$. In general,

Lemma 3.1. *A type $(1,1)$ quantifier is CONSERV and EXT iff it is the relativization of a type (1) quantifier.*

Definition 3.3.4 Let $\mathcal{L}$ be a logic, and Q a quantifier. We say that $\mathcal{L}(Q)$ has the *relativization property* (or that Q relativizes with respect to $\mathcal{L}$) if Q^r is definable in $\mathcal{L}(Q)$, i.e. if $\mathcal{L}(Q^r) \le \mathcal{L}(Q)$.

An alternative definition can be given in terms of structures:

Definition 3.3.5 Let P be a unary predicate, $\mathcal{M} =< A, R_1,\ldots,R_m >$ a structure, and define $\mathcal{M} \uparrow P$ as the structure $< P, R_1 \uparrow P,\ldots,R_m \uparrow P >$. Then $\mathcal{L}(Q)$ has the *relativization property* if for any $\mathcal{M}$, P, and sentence $\phi \in \mathcal{L}(Q)$ the relativization of ϕ is the sentence ϕ^P such that $\mathcal{M} \models \phi^P$ iff $\mathcal{M} \uparrow P \models \phi$.

---

[8] We suppress the assignment to variables not displayed.

For an example of a quantifier that does *not* relativize, take the quantifier **more**, which is the relativization of the Rescher quantifier, $\mathbf{Q_R}$. We have that **more** is not expressible in $\mathcal{L}(\mathbf{Q_R})$. Another example is **most**, which is not definable from FOL even when supplemented with the unary **most**[1] (which can be paraphrased as **more than half of all things**); this is true even when we restrict ourselves to finite models ([16]). Westerståhl extended this result to all quantifiers that we called *proportional* ([104]). But the more general result of this kind is probably the following:

Theorem 3.3.6 ([65]) Let $\mathcal{L}$ be a logic. If a quantifier Q follows EXT, then $\mathcal{L}(Q)$ relativizes.

This theorem tells us that $\mathbf{Q_R}$ and **most** is *not* domain independent, which is not difficult to see. In fact, [65] tells us a stronger result: **most**[(1,1)] is not definable even in *infinitary logic* extended with a finite set of type (1) quantifiers[9]. Again, this is true even when we restrict ourselves to finite models.

Another relevant definition is that of lifting.

Definition 3.3.7 If Q is a quantifier of type $(k_1, \ldots, k_m)$, and $n > arity(Q)$ a natural number, the lifting of Q to n (written $Q^{(n)}$) is the quantifier of type $(nk_1, \ldots, nk_m)$ defined by, for all M and all $A_i \subseteq M^{nk_i}$ ($1 \le i \le m$),

$$Q_M^{(n)}(A_1, \ldots, A_k) \text{ iff } Q_{M^n}(A_1, \ldots, A_k)$$

The lifting mechanism allows us to apply a quantifier defined in a domain M to sets made up of tuples of elements of M (pairs, triples, etc.). Note, though, that we consider the tuples as basic, and do not take into account their internal structure. As an example, $\mathbf{all}^{(2,2)} = \{A, B \subseteq M^2 \mid A \subseteq B\}$ checks to see if a binary relation is a subset of another one. This does not allow us to look "inside" the relations A or B and determine their properties; $\mathbf{Q_T}^{(2)} = \{A \subseteq M^2 \mid A \text{ is transitive}\}$ is not the lifting of any quantifier of type (1) (or of any set of such quantifiers).

Theorem 3.3.8 ([62]) If Q satisfies ISOM (EXT, CONS) then so does $Q^{(n)}$.

Definition 3.3.9 The *Boolean relatives* of quantifier Q are defined as one of the following:

- Let Q be a quantifier of type $[k_1, \ldots, k_m]$. Then the *negation* of Q on M, in symbols Q_M^{-}, is a quantifier of type $[k_1, \ldots, k_m]$ defined by, for all M, all $A_i \subseteq M^{K_i}$ ($1 \le i \le m$)
  $Q_M^{-}(A_1, \ldots, A_m)$ iff it is not the case that $Q_M(A_1, \ldots, A_m)$.

---

[9] Infinitary logic is an extension of FOL in which *infinite* conjunctions and disjunctions (that is, conjunctions and disjunctions with an infinite (numerable) number of conjuncts or disjuncts) are allowed. Such a logic is extremely powerful.

- Let Q_1, Q_2 be boolean relatives of Q, both of type $[k_1, \ldots, k_m]$. Then the *conjunction* of Q_1 and Q_2 is the quantifier $Q^\wedge$ of type $[k_1, \ldots, k_m]$ defined by, for all M, all $A_i \subseteq M^{K_i}$ $(1 \leq i \leq m)$
  $$Q^\wedge_M(A_1, \ldots, A_m) \text{ iff } (Q_1)_M(A_1, \ldots, A_m) \text{ and } (Q_2)_M(A_1, \ldots, A_m).$$
- Let Q_1, Q_2 be boolean relatives of Q, both of type $[k_1, \ldots, k_m]$. Then the *disjunction* of Q_1 and Q_2 is the quantifier $Q^\vee$ of type $[k_1, \ldots, k_m]$ defined by for all M, all $A_i \subseteq M^{K_i}$ $(1 \leq i \leq m)$
  $$Q^\vee_M(A_1, \ldots, A_m) \text{ iff } (Q_1)_M(A_1, \ldots, A_m) \text{ or } (Q_1)_M(A_1, \ldots, A_m).$$

All the Boolean relatives and liftings of a quantifier Q are closely related to the structural property that Q expresses, so we would like to close quantifiers under all the properties above. A similar idea is used in [24], where the set of all liftings of a GQ Q is called the *uniform sequence* generated by Q. We say that Q is *atomically reducible* to Q' if Q is definable from Q' with atomic (quantifier-free) FOL formulas ([80]). We then have that, for many natural classes of quantifiers $\mathcal{Q}$, if $Q \in \mathcal{Q}$, then $Q^{(n)}$ and any Boolean relative of Q are atomically reducible to Q. This motivates the following definition:

Definition 3.3.10 The *family* of a given GQ Q (written $Fam(Q)$) is defined inductively as

- $Q \in Fam(Q)$.
- If $Q \in Fam(Q)$, then for any n $Q^{(n)} \in Fam(Q)$.
- If $Q \in Fam(Q)$, then any Boolean relative of Q is in $Fam(Q)$.

It is then possible to define $\mathcal{L}(Q)$, for any GQ Q and logic $\mathcal{L}$, as the result of adding $Fam(Q)$ to $\mathcal{L}$. This process is well defined, as theorem 3.3.8 and the above observation by Otto ensure us that the added quantifiers all behave uniformly, and the resulting logic is also well behaved. The following result just confirms this:

Theorem 3.3.11 If quantifier Q is definable on $\mathcal{L}$, $Fam(Q)$ is definable in $\mathcal{L}$.

3.4 Basic Complexity

The question on which most research is centered is that of *expressive power*. The most basic question is simply: given a logic L and GQ Q, what properties are expressible in $L(Q)$ that are not expressible in L? From here, theory attempts to answer more general questions: given GQs Q and Q', how do $L(Q)$ and $L(Q')$ compare? Or, given two sets of quantifiers $\mathcal{Q}$ and $\mathcal{Q}'$, how do $L(\mathcal{Q})$ and $L(\mathcal{Q}')$ compare?

Here we overview only a few of the basic results. To put these results in perspective, we start with the most general one. We have already seen above that families of quantifiers do not increase their expressive power; that is essentially the last theorem's content. Beyond families, however, the expressive power of GQs increases with its type. Hella proved the basic result that created a *hierarchy* of GQs, as follows: we can define a *lexicographic* order on types:

$$(1) < (1,1) < \ldots < (2) < (2,1) < (2,1,1) < \ldots < (2,2) < \ldots (3) < \ldots$$

Then for every type t there is a quantifier of that type that is not definable in $FOL(\mathcal{Q})$ where all quantifiers in $\mathcal{Q}$ are of some type $t' < t$ ([47, 48]).

Thus, given any set of GQs $\mathcal{Q}$ that has bounded arity (i.e. there is an n such that the arity of all GQs in the set is $\leq n$), there is a property of $n+1$-ary relations (equivalently, a class of structures with a signature that includes some $n+1$ relation name) that is not definable in $FOL(\mathcal{Q})$. In fact, the property can be a PTIME property.

One corollary of this result was to answer, in the negative, a conjecture[10] that stated that PTIME could be captured by a logic extended with the appropriate set of Generalized Quantifiers. No such finite set of GQs exist.

Even though as a class monadic quantifiers are the simplest ones, there are differences among quantifiers in the class. The following are a few facts about the expressive power of some common monadic quantifiers.

Lemma 3.2. *([102])*

-
$$FOL < \left\{ \begin{array}{ll} L(Q_\omega) < & L(I) \\ L(Q_R) < & L(most) \end{array} \right\} < L(more) < L(H)$$

  Logics on different branches are incomparable, but each logic is strictly stronger than the ones preceding it.
- $L(Q_E) \not\leq L(more)$.
- *In general, $L(most) < L(more)$, but $L(most) = L(more)$ on finite structures.*
- $L(most, Q_\omega) = L(more)$.

In the above, L is taken to be a logic at least as powerful as FOL. It turns out that for more powerful, but practical logics the same results hold. Let $FIXPOINT$ be the logic that results from adding a *least fixpoint* operator to FOL, FOC the result of adding *counters* and basic arithmetic to FOL, and $WHILE$ the result of adding unrestricted looping ability to FOL. GQs can be added to any of these languages.

Definition 3.4.1 Let $\mathcal{Q}$ be a set of generalized quantifiers. Then $FOL(\mathcal{Q})$ $(FIXPOINT(\mathcal{Q}), WHILE(\mathcal{Q}), FOC(\mathcal{Q}))$ is the logic obtained by adding $\mathcal{Q}$ to FOL ($FIXPOINT, WHILE, FOC$) by means of corresponding syntactic and semantic rules.

Adding a finite number of quantifiers to query languages does not change the basic hierarchy:

---

[10] It seems that this conjecture was first posed by Immerman.

Theorem 3.4.2 For any finite set of quantifiers $\mathcal{Q}$,

$$FOL(\mathcal{Q}) \subseteq FIXPOINT(\mathcal{Q}) \subseteq WHILE(\mathcal{Q})$$

However, the exact expressiveness of the language obtained depends on the kind of quantifiers that we add, specifically on the *arity* of the quantifier and the basic type of property (invariant) that it expresses.

Clearly, type (1) quantifiers are the most basic of all, followed by type $(1,1)$ quantifiers. And yet, we will focus our efforts on later chapters in these types of quantifiers. Why do we use only the most basic (and limited) types of quantifiers in our efforts? The reason has to do with the basic goal of the work presented here: to come up with *practical* query languages, that is, languages which fulfill at least two requisites:

- they can express commonly posed queries, as opposed to esoteric properties that rarely (if ever) would appear in an application; and
- they can be implemented efficiently by a suitable query processor and optimizer.

The first condition is admittedly vague and pragmatic, so it's doubtful that it admits proper formalization. The second one, however, can be given a formal treatment: efficient implementation in databases can be identified with problems that are within the PTIME complexity class -and, in practice (with the help of indices and other devices) can be made to run in $O(n \log n)$ or very close to it[11]. Thus, one constraint is to restrict ourselves to *low complexity* quantifiers. As we will see, many *monadic* quantifiers fulfill this requirement. At the same time, though, the GQs have to be *useful*, however we chose to define this property. In fact, one of the main points of our research is to identify a class of quantifiers that is of practical use *and* can be implemented efficiently. From among the monadic quantifiers, we argue that those of type (1) and $(1,1)$ are some of the most practical ones. While, as stated above, this is a largely pragmatic argument, there are well founded reasons to support this viewpoints; some of them will be introduced in chapter 8.

What about *polyadic* quantifiers? Can't an argument be made that they are also quite useful? Certainly they can be. Consider, however, the simplest polyadic class: type (2). Such quantifiers can be identified with *graph properties* -since their argument is a binary relation. This class already includes well known NP problems: deciding whether an arbitrary graph has a Hamiltonian circuit[12] is one of the best known ones. Thus, the quantifier $\mathbf{Q_{Ham}}$, introduced earlier, is not one that we want to include in a query language. The problem, then, is to come up with a subclass of type (2) quantifiers that has low complexity and still offers practical applications. Unfortunately, such

---

[11] The reason for tighter practical bounds is that, in databases, n tends to be pretty large, and even polynomial time may be too much for some situations.

[12] Also called a Hamiltonian cycle.

a subclass has not been identified yet. Thus, in this work we limit ourselves to the above mentioned monadic quantifiers.

Note that some individual quantifiers (of arbitrary type) may be identified that have low complexity. However, the aim is to add, to a query language, a *regular* set of quantifiers, that is, a set such that all the members can be treated in a uniform manner. Thus, creating a set by adding arbitrary quantifiers to a query language is not considered a good approach. The reason is that such procedure will produce a *finite* set of quantifiers[13], each one of them requiring its own implementation as a separate operator. Such set is unlikely to compensate for the work required to support it with a corresponding payoff in terms of the queries that are now expressible by the quantifiers in the set. Thus, a more uniform approach is called for, and that is what we will try to develop in later chapters.

---

[13] We assume that methods like adding the Boolean family of a quantifier are not used, since they do not really add to the expressive power of the language.

4

$QLGQ$: A Query Language with Generalized Quantifiers

4.1 Introduction: GQs in query languages

The idea of using GQs in query languages can be traced back to the developments of extensions of relational algebra that hinge on the idea that *dealing with sets* is a necessary tool for a query language. SQL contains a rudimentary version of generalized quantification in the EXISTS predicate[1]. However, SQL's syntax does not allow for any other kind of quantification except universal quantification which is achieved by combining negation and existential quantification (the NON EXISTS predicate).

There have been three kinds of proposals that are directly relevant to the present work. The first one is the incorporation of extended quantification, in the form of a *universal quantifier*, to the relational algebra. The second one is the incorporation of set predicates to the algebra. The third one is a direct attempt to incorporate Generalized Quantifiers to SQL. All these proposals work only at the language level, adding new constructs to the query language. Other proposals work at the implementation level, proposing more efficient support for already existing constructs.

Our motivation for creating a new query language is to be able to study issues of generalized quantification in a setting as general as possible. To this end, the language introduced is tailored to the use of GQs, while being kept as simple as possible. We will also study the issue of how to use GQs in SQL; we will see then that there are several choices one can make with respect to the exact syntax required. However, we don't want such choices to interfere with the analysis of quantification. Therefore, in the next section we introduce the language that we will use throughout the rest of the book.

---

[1] Technically, this predicate corresponds to an infinite class of quantifiers, namely the class $\{\mathbf{some}^n \mid n > 0\}$.

A.Badia, *Quantifiers in Action: Generalized Quantification in Query. Logical and Natural Languages*, Advances in Database Systems 37, DOI: 10.1007/978-0-387-09564-6_4,
© Springer Science+Business Media, LLC 2009

4.2 *QLGQ*

In this section, we introduce *QLGQ* (Query Language with Generalized Quantifiers). We will show, together with a formal description of the language, a simple example. Let us imagine a simple domain of discourse, a universe of people and some predicates that hold on the domain elements. We will use a language made up of variables, constants, predicate and quantifier symbols (which roughly correspond to natural language determiners like *every, two, at least one, some, most . . .*) to write sentences about this world.

In order to define the query language and its semantics, we will first introduce the idea of *vocabularies* and *interpretations*. As previously, we consider a set $\mathcal{U}$ as the fixed universe.

Definition 4.2.1 A *GQ-vocabulary* γ is a triple $[(\mathcal{R}, arity), \mathcal{C}, (\mathcal{Q}, type)]$ where

1. $\mathcal{R}$ is a finite set of *predicate names* and *arity* is a *relation arity function*, $arity : \mathcal{R} \rightarrow \mathcal{N}$ which associates with each predicate name $R \in \mathcal{R}$ its arity $arity(R)$;
2. $\mathcal{C}$ is an enumerable set of *constant names*; and
3. $\mathcal{Q}$ is an enumerable set of *quantifier names* and *type* is a *quantifier type function*, $type : \mathcal{Q} \rightarrow \mathcal{N}^{\mathcal{N}}$, which associates with each quantifier name $Q \in \mathcal{Q}$ its type $type(Q)$.

Example 4.1. Consider the GQ-vocabulary
$$\gamma_1 = [(\{\texttt{EMP}, \texttt{WORKS-ON}\}, arity), \{\texttt{Johnson}, \texttt{Smith}\}, (\{\textbf{some}, \textbf{no}, \textbf{at least 2}, \textbf{all}\}, type)]$$
where $arity(\texttt{EMP}) = arity(\texttt{WORKS} - \texttt{ON}) = 2$, *and*
$type(\textbf{some}) = type(\textbf{no}) = type(\textbf{at least 2}) = type(\textbf{all}) = (1, 1)$.

We assume that the first (second) attribute of the predicate name $\texttt{EMP}$ *corresponds to the social-security number (name) of an employee, and we assume that the first attribute of the predicate name* $\texttt{WORKS-ON}$ *corresponds to the social-security number of an employee and the second attribute corresponds to the name of a project that the employee works on.*

Definition 4.2.2 Let $\gamma = [(\mathcal{R}, arity), \mathcal{C}, (\mathcal{Q}, type)]$ be a *GQ-vocabulary*. A γ-*interpretation* D is a tuple $(D_{\mathcal{R}}, D_{\mathcal{C}}, D_{\mathcal{Q}})$, where $D_{\mathcal{R}}$ is a function which associates with each relation name $R \in \mathcal{R}$ a finite subset of $\mathcal{U}^{arity(R)}$, $D_{\mathcal{C}}$ is a function that associates with each constant in $\mathcal{C}$ an element in $\mathcal{U}$, and $M_{\mathcal{Q}}$ is a function which associates with each quantifier name $Q \in \mathcal{Q}$, such that $type(Q) = (k_1, \ldots, k_n)$, a n-ary relation over sets $A_1, \ldots, A_n$, where A_i is an k_i-ary relations over $\mathcal{U}$. In addition, we require that $D_{\mathcal{C}}$ be an *injective* mapping.

Given a *GQ-vocabulary* $\gamma = [(\mathcal{R}, arity), \mathcal{C}, (\mathcal{Q}, type)]$, we call $[(\mathcal{R}, arity), \mathcal{C}]$ the *relational part* of the vocabulary; such a part defines a *database schema*, with attributes identified by position instead of by name. Likewise, given a γ-interpretation $D = (D_{\mathcal{R}}, D_{\mathcal{C}}, D_{\mathcal{Q}})$, we call $(D_{\mathcal{R}}, D_{\mathcal{C}})$ the *relational part* of the interpretation and note that it defines a *database instance*. At some points we will consider some set $D_{\mathcal{Q}}$ as fixed, and identify a relational database D with the γ interpretation D.

A *query* is a generic, computable mapping between two database instances ([6]). For a language $\mathcal{L}$ and γ-interpretation D, a formula φ with free variables $x_1, \ldots, x_n$ (or, alternatively, a set term $\{x_1, \ldots, x_n \mid \varphi\}$) denote a query, $[\![\varphi]\!]_D$, as defined by the semantics of $\mathcal{L}$ with respect to interpretation D. Given two query languages $\mathcal{L}$ and $\mathcal{L}'$, we say that a formula $\varphi \in \mathcal{L}$ and formula $\psi \in \mathcal{L}'$ are *equivalent* if, for all interpretations D, $[\![\varphi]\!]_D = [\![\psi]\!]_D$.

4.2.1 Syntax of *QLGQ*

Here we introduce the syntax of our query language.

Definition 4.2.3 The expressions φ in the language $QLGQ(\mathcal{Q})$ and the set of free variables (in symbols, $Fvar(\varphi)$) is defined as follows:

1. *Basic terms*
   a) If $x \in \mathcal{V}$ then x is a basic term. $Fvar(x) = \{x\}$
   b) If $c \in \mathcal{C}$ then c is a basic term. $Fvar(c) = \emptyset$
2. *Set terms*
   a) If φ is a formula, and $[x_1, \ldots, x_m]$ is a multiset (i.e. repetitions are allowed), and $\{x_1, \ldots, x_m\} \subseteq Fvar(\varphi)$, then $\{x_1, \ldots, x_m \mid \varphi\}$ is a set term. The set $\{x_1, \ldots, x_m\}$ is called the *selector* of the set term (in symbols, $Sel(\{x_1, \ldots, x_m \mid \varphi\})$).
   $Fvar(\{x_1, \ldots, x_m \mid \varphi\}) = Fvar(\varphi) - \{x_1, \ldots, x_m\}$.
   b) If X_1, X_2 are set terms, then $X_1 \cup X_2$ is a set term, provided that $Fvar(X_1) = Fvar(X_2)$.
   $Fvar(X_1 \cup X_2) = Fvar(X_1) \cup Fvar(X_2)$.
3. Formulas:
   a) If t_1, t_2 are basic terms, and θ is a comparison operator[2]. then $t_1 \theta\, t_2$ is a formula.
   $Fvar(t_1 \theta\, t_2) = Fvar(t_1) \cup Fvar(t_2)$.
   b) If $R \in \mathcal{R}$, arity of R is n and $t_1, \ldots, t_n$ are basic or anonymous terms, then $R(t_1, \ldots, t_n)$ is a (basic) formula.
   $Fvar(R(t_1, \ldots, t_n)) = Fvar(t_1) \cup \ldots \cup Fvar(t_n)$.
   c) If φ_1 is a formula and φ_2 is a formula, then $\varphi_1 \wedge \varphi_2$ is a formula.
   $Fvar(\varphi_1 \wedge \varphi_2) = Fvar(\varphi_1) \cup Fvar(\varphi_2)$.
   d) If $Q \in \mathcal{Q}$ is a quantifier name of type $[k_1, \ldots, k_m]$ and $X_1, \ldots, X_m$ are set terms with $\text{arity}(X_i) = k_i$ for $1 \leq i \leq m$, then $Q(X_1, \ldots, X_m)$ is a formula.
   $Fvar(Q(X_1, \ldots, X_m)) = Fvar(X_1) \cup \ldots \cup Fvar(X_m)$.

---

[2] It is customary to consider that θ can be any of $=, \neq, \leq, \geq, <, >$, and assume that an operator is only used where it makes sense (i.e. in ordered domains, and so on).

A set term X with free variables (that is, $Fvar(X) \neq \emptyset$) is called *parametric*; one without, *non-parametric*. Any free variables are called *parameters*; and the non-free variables, *bounded*. A *query* is a non-parametric set term. A *sentence* is a formula φ such that $Fvar(\varphi) = \emptyset$.

Example 4.2. Before formalizing the semantics of the language, we give a few intuitive examples. In the following, we assume a database with relations: **Student(sid)** and **Professor(pid)**, with the obvious meaning; **Takes(sid,cid)**, which denotes that student **sid** is enrolled in course **cid**, **Lectures(pid,cid)**, which denotes that professor **pid** gives the lectures for course **cid**; **Teaches(pid,sid)**, which denotes the professor **pid** is a teacher of student **sid**; and **Friend(sid1,sid2)** which holds between two students if the first is a friend of the second.

1. Find the professors teaching all students.

$$\{y \mid \mathbf{all}\{x \mid \text{Student}(x)\}, \{x \mid \text{Teaches}(y, x)\}\}$$

2. Find the professors teaching no students.

$$\{y \mid \mathbf{no}\{x \mid \text{Student}(x)\}, \{x \mid \text{Teaches}(y, x)\}\}$$

3. Find the professors teaching all but one of the students.

$$\{y \mid \mathbf{all\ but\ one}\{x \mid \text{Student}(x)\}, \{x \mid \text{Teaches}(y, x)\}\}$$

4. Find the professors teaching at least two students.

$$\{y \mid \mathbf{at\ least\ two}\{x \mid \text{Student}(x)\}, \{x \mid \text{Teaches}(y, x)\}\}$$

5. Find the professors teaching half of the students.

$$\{y \mid \mathbf{half}\{x \mid \text{Student}(x)\}, \{x \mid \text{Teaches}(y, x)\}\}$$

6. Find the pairs of teachers teaching the same number of students.

$$\{x, y \mid \mathbf{I}(\{z \mid \text{Teaches}(x, z)\}\}, \{z \mid \text{Teaches}(y, z)\}\})$$

In the context of *QLGQ*, a query is simply a set term. Since set terms can be nested, complex queries can be expressed by combining different GQs. As an example, the query *"List the students who have at least 2 friends taught by Peter in some course"* can be expressed in *QLGQ* as follows:

$$\{Sname \mid \mathbf{at\ least\ two}$$
$$(\{fr \mid \text{Friend}\ (Sname, fr)\},$$
$$\{fr \mid \mathbf{some}$$
$$(\{crs \mid \text{Lectures}(\mathbf{Peter}, crs)\},$$
$$\{course \mid \text{Takes}(fr, course)\})\})\}$$

4.2.2 Semantics of *QLGQ*

We are now ready to define the *semantics function* ($\llbracket \ \rrbracket$) for terms and the *satisfaction relation* ($\models$) for formulas of $QLGQ(\gamma)$. We point out that, as the definition of terms and formulas in $QLGQ(\gamma)$ is mutually recursive, so are the semantics function and the satisfaction relation, which are defined in terms of each other[3].

Definition 4.2.4 Let D be a γ-interpretation.

1. *Basic terms*
   a) For $a \in \mathcal{U}$, then $\llbracket x \rrbracket_{D,a} = a$.
   b) $\llbracket c \rrbracket_D = D_C(c)$.
2. *Set terms*
   a) For $a_{m+1}, \ldots, a_n \in \mathcal{U}$,

   $$\llbracket \{x_1, \ldots, x_m \mid \varphi(x_1, \ldots, x_n)\} \rrbracket_{D,a_{m+1},\ldots,a_n} =$$

   $$\{(a_1, \ldots, a_m) \in \mathcal{U}^m \mid D, a_1, \ldots, a_n \models \varphi\}.$$

   b) For $a_{m+1}, \ldots, a_n \in \mathcal{U}$,

   $$\llbracket S_1 \cup S_2 \rrbracket_{D,a_{m+1},\ldots,a_n} =$$

   $$\llbracket S_1 \rrbracket_{D,a_{m+1},\ldots,a_n} \cup \llbracket S_2 \rrbracket_{D,a_{m+1},\ldots,a_n}$$

3. *Formulas*
   a) In order to give proper semantics to comparisons, it is necessary to proceed with a case analysis. Since a basic term may be a constant or a variable, we have the following cases[4]:
      i. For $a_1, a_2 \in \mathcal{U}$
      $$D, a_1, a_2 \models x_1 \theta\ x_2$$
      if $\llbracket x_1 \rrbracket_{D,a_1}\ \theta\ \llbracket x_2 \rrbracket_{D,a_2}$.
      ii. For $a \in \mathcal{U}$
      $$D, a \models x \theta x$$
      if $\llbracket x \rrbracket_{D,a}\ \theta\ \llbracket x \rrbracket_{D,a}$.
      iii. For $a \in \mathcal{U}$
      $$D, a \models x \theta c$$
      if $a \theta \llbracket c \rrbracket_D$.
      iv.
      $$D \models c_1 \theta c_2$$
      if $\llbracket c_1 \rrbracket_D\ \theta\ \llbracket c_2 \rrbracket_D$.

---

[3] For simplicity of presentation, we define the language only for *regular* quantifiers (see section 3.2). In the present work we will rarely need *irregular* quantifiers; but adding them to the language presents no technical difficulties.

[4] We disregard symmetric cases and assume that comparison operator θ evaluates to itself.

b) Let $n = |Fvar(R(t_1, \ldots, t_{arity(R)}))|$. Then

$$D, a_1, \ldots, a_n \models R(t_1, \ldots, t_{arity(R)})$$

if $(b_1, \ldots, b_{arity(R)}) \in D_{\mathcal{R}}(R)$, where $b_i = a_j$ $(1 \le i \le arity(R)$, $1 \le j \le n)$ if t_i is the variable x_j, or $b_i = [\![t_i]\!]_D$, $1 \le i \le arity(R)$, if t_i is a constant.

c) Let $Fvar(\varphi_1) \cup Fvar(\varphi_2) = \{x_1, \ldots, x_n\}$, and let $\bar{a} = (a_1, \ldots, a_n)$. Define $\bar{a}_{\varphi_1} = (a_i)_{x_i \in Fvar(\varphi_1)}$, and similarly for $\bar{a}_{\varphi_2}$. Then $D, \bar{a} \models \varphi_1 \wedge \varphi_2$ if $D, \bar{a}_{\varphi_1} \models \varphi_1$ and $D, \bar{a}_{\varphi_2} \models \varphi_2$.

d) Let $Fvar(S_1) \cup \ldots \cup Fvar(S_k) = \{x_1, \ldots, x_n\}$, and let $\bar{a} = (a_1, \ldots, a_n) \in A^n$. Define, as before, $\bar{a}_{S_i}$ as $(a_i)_{x_i \in Fvar(S_i)}$, for $1 \le i \le k$. Then $D, \bar{a} \models Q(S_1, \ldots, S_k)$ if

$$([\![S_1]\!]_{D, \bar{a}_{S_1}}, \ldots, [\![S_k]\!]_{D, \bar{a}_{S_k}}) \in D_{\mathcal{Q}}(Q)$$

4.2.3 Remarks on Syntax

It is worth remarking some characteristics of the language which may not obvious at first glance. First and foremost, we must acknowledge that what we have just defined is the *family* of languages *QLGQ*, not *a* language. We will obtain concrete query languages from it by specifying a set of generalized quantifiers to be added to the basic language. Note that rule 3d of the language syntax can only be used when some quantifier is available. The language with no quantifiers, which we will write simply *QLGQ*, must be understood as the language without rule 3d in its syntax or semantics. One of the main areas of our research will be to find out which quantifiers we must add to *QLGQ* to make it an interesting query language.

Clearly, the basic formulas in *QLGQ* correspond to SPJ (Select-Project-Join) queries in Relational Algebra[5]. As for union, it can be represented with union of set terms. But it is interesting to note that *QLGQ* does not use $\neg$, $\vee$, or the quantifiers $\exists, \forall$ as formula constructors. This is in contrast with *FOL* or the language $L(GQ)$ of Barwise and Cooper ([16]), where these constructors are used. However, the language still keeps the natural flavor of $L(GQ)$, in the sense that most queries are formulated in *QLGQ* in a manner that resembles the way in which they would be formulated in natural language. Note, in particular, queries 5, 6 and 7 of example 4.2, in which the only thing that changes in the natural language formulation of the queries is the determiner, and correspondingly the only thing that changes in the *QLGQ* queries is the quantifier used. And yet, it can be seen that in combination with a reasonable set of generalized quantifiers the language has the expressive power of *FOL*. These observations lead to the proposition of the *conjunctive formulation thesis* ([12]):

---

[5] Technically, Cartesian product is the primitive operator -and can indeed be expressed in basic *QLGQ* thanks to conjunction.

*The natural way to formulate a real-world query is as a conjunction
of statements. These statements may either be*

- *simple first order predicate statements; or*
- *set predicate statements over potentially complex sub-queries.*

In particular, it is interesting to point out that the missing first order constructions will be incorporated into the language via quantifiers. The next theorem shows how:

Theorem 4.2.5 Every *FOL* query is expressible as a *QLGQ*(**some**, **no**) query.

Another point of interest is the relationship between this language and relational algebra. It may seem, in light of the previous theorem, that the relationship is obvious, as *FOL* and relational algebra are considered roughly equivalent. That "roughly", however, hides a series of technical problems that the presence of generalized quantification makes necessary to deal with. It is well known that relational algebra is a safe and domain independent language (basically, it never returns infinite answers or answers that depend on elements outside relations). This is not so in *FOL*; thus, relational algebra is equivalent to the safe, domain independent fragment of *FOL*. Unfortunately, there is no positive characterization of such fragment, although there are ways to limit *FOL* to only safe expressions. Thus, we need to find a safe (and, if possible, domain independent) fragment of *QLGQ*. But even after that is done, there is a mismatch between the relational and logical approaches, which prevents a *direct* translation of *QLGQ* formulas to relational algebra. Since such a direct translation is needed for our strategy to implement *QLGQ* queries efficiently, we will also have to deal with the mismatch. Thus, we devote the next sections and part of next chapter to address these issues.

Before we tackle the problems, we add a bit of *syntactic sugar* to the language. In order to make expression of some queries simpler, we will introduce the following two shortcuts:

- '_' is a term, the *anonymous* term. $Fvar(\_) = \emptyset$
- For formula φ, if $\{x_1, \ldots, x_n\} \subseteq Fvar(\phi)$ and $\{y_1, \ldots, y_m) \subseteq Fvar(\varphi)$, with $\{x_1, \ldots, x_n\} \cap \{y_1 \ldots, y_m\} = \emptyset$,

$$\{x_1, \ldots, x_n, (y_1, \ldots, y_m) \mid \varphi\}$$

is a set term α, and

$$Fvar(\alpha) = Fvar(\varphi) - \{x_1, \ldots, x_n\} \cup \{y_1 \ldots, y_m\}$$

The motivation behind both rules is the same: in *QLGQ*, we don't admit vacuous use of variables. Note that in *FOL* the formula $\exists x\ \psi$ can always be used, whether x appears among the variables of ψ or not. If x does not appear at all in ψ or if it appears but it is bound, this formula is equivalent to ψ. But in *QLGQ* variables are used to be captured (in which case they are used for quantification, as in FOL, or they are to be parameters. This creates a problems with simple queries like

$$\{x \mid R(x, y)\}$$

or

$$\{x \mid R(x, y) \wedge y > 5\}$$

where R is a binary relation: y becomes a parameter, whether we intend that or not. It must be used because of the syntax of the language. If we really meant to project the relation onto the first attribute, then we will use from now on:

$$\{x \mid R(x, \_)\}$$

or

$$\{x(y) \mid R(x, y) \wedge y > 5\}$$

In both cases, the set terms are *not* parametric, and the intended meaning is that of projection. Note that this is simply an application of existential quantification (the quantifier **some**[1]), so this is available to us in any language where such quantifier is present -and it will be in most cases.

4.3 Safety and Domain Independence

We now study how changes to the domain of interpretation affect the semantics of queries. For the purposes of the research presented here it is necessary to find out how (if) the query denoted by a formula changes when the underlying domain used to interpret the formula changes. Thus, we will make the domain of interpretation explicit as follows. For any relation R, set $M \subseteq \mathcal{U}$, $R \uparrow M = \{(a_1 \ldots, a_{arity(R)}) \in R \mid (a_1, \ldots, a_{arity(R)}) \in M^{arity(R)}\}$. Given a relation $R \in \mathcal{U}^{arity(R)}$, $adom(R) = \{a_i \mid (a_1, \ldots, a_{arity(R)}) \in R \wedge 1 \leq i \leq arity(R)\}$. For interpretation $D = (D_\mathcal{R}, D_\mathcal{C}, D_\mathcal{Q})$, $adom(D) = \bigcup_{R \in D_\mathcal{R}} adom(R) \cup D_\mathcal{C}$ is the *active domain* of the database instance denoted by D. For database D, any set M such that $adom(D) \subseteq M \subseteq \mathcal{U}$, we will define the semantics of query languages against the *interpretation generated by M*, $D_M = (M, D_\mathcal{R} \uparrow M, D_\mathcal{C} \uparrow M, D_\mathcal{Q} \uparrow M)$, where $D_\mathcal{R} \uparrow M = \{R \uparrow M \mid R \in D_\mathcal{R}\}$ and $D_\mathcal{Q} \uparrow M = \{Q \uparrow M \mid Q \in D_\mathcal{Q}\}$. The set M is called the *domain of evaluation*. Note that since $adom(D) \subseteq M$, $D_\mathcal{R} \uparrow M = D_\mathcal{R}$ and $D_\mathcal{C} \uparrow M = D_\mathcal{C}$. Thus, we will write D_M simply as $(M, D_\mathcal{R}, D_\mathcal{C}, D_\mathcal{Q} \uparrow M)$. In particular, interpretation $(adom(D), D_\mathcal{R}, D_\mathcal{C}, D_\mathcal{Q} \uparrow adom(D))$ gives the *active domain* semantics of a language, and interpretation $(\mathcal{U}, D_\mathcal{R}, D_\mathcal{C}, D_\mathcal{Q})$ gives the *natural semantics* of a language. Note that for any query φ, any interpretation D_M, it is the case that $[\![\varphi]\!]_{D_M} \subseteq M^n$, where n is the number of free variables in φ.

As Kifer points out ([64]) there is confusion in the literature on the use of the terms *safe* and *domain independent*. Since our interest is in the properties themselves, our safe corresponds to *semantically* safe in the literature, and to *universally* safe in [64]. The same holds for the concept of *domain independent*.

Definition 4.3. A query φ is *safe* if for any interpretation D, any set M such that $adom(D) \subseteq M \subseteq \mathcal{U}$, $[\![\varphi]\!]_M$ is finite.

Informally, a query is safe if it does not return an infinite answer under any (infinite) domain of evaluation.

Definition 4.4. A query φ is *domain independent* if for any interpretation D, any two sets M, M' such that $adom(D) \subseteq M$, $M' \subseteq \mathcal{U}$, $[\![\varphi]\!]_M = [\![\varphi]\!]_{M'}$.

In other words, the answer to φ is the same regardless of the domain of evaluation.

If all the queries in a language $\mathcal{L}$ are safe (domain independent) we say that the language is safe (domain independent, strongly safe).

These concepts are related:

Lemma 4.5. *If a query φ is domain independent, then φ is safe.*

The converse of the above proposition does not hold. This can be shown by giving an example of a query that is safe but not domain independent.

Example 4.6. Such an example is provided in [6]:

$$\{x|\ \mathbf{all}^1\{y\ |\ R(x,y)\}\}^6$$

Here R is a relation on the database. Since all values of x come from R, the number of values for x is finite and the query is safe. However, if we vary our universe, the answer to the query changes, since which x qualifies depends on a relation with all the elements in the universe. Note that since we assume that relations are finite, the answer to this query can only be non empty if the universe of discourse is finite too. In the natural semantics, the query always evaluates to the empty set; in the active domain semantics, the evaluation depends on the extension of R (and other relations in the database), but it is easy to build databases in which the answer is not empty.

We would like to determine what is the behavior of $QLGQ$ with respect to safety and domain independence with a given semantics. For the standard case, the answer is simple:

Lemma 4.7. *Let Q be an arbitrary set of GQs. Then $QLGQ(Q)$ (under the natural semantics) is an unsafe language.*

It is pretty obvious that the set $\{x|x\ =\ x\}$ is an unsafe set. But even if we restrict the use of equality, there is a more subtle problem with the language as described. It is perhaps better to show the problem through an example. Consider a GQ-vocabulary $\gamma_2\ =\ [(\{R,S\},\alpha_2),\emptyset,(\{\mathbf{all}\}),\beta_2]$, with $\alpha_2(R)\ =\ 2$ and $\alpha_2(S)\ =\ 1$, and $\beta_2(\mathbf{all})\ =\ [2,1]$, and the query:

---

[6] For all $A \subseteq \mathcal{U}$, $\mathbf{all}^1{}_M(A)\ =\ \{A\}$.

$$\{x \mid \textbf{all}$$
$$(\{y \mid \mathrm{R}(y,x)\})$$
$$(\{z \mid \mathrm{S}(z)\})\}$$

We analyze its meaning in a given γ_2-interpretation $M = [A, M_r, M_Q]$, and assume for the moment, the natural semantics. Then,

$$[\![(\{x \mid \textbf{all}(\{y \mid \mathrm{R}(y,x)\}\,\{z \mid \mathrm{S}(z)\})\})]\!]_M =$$
$$\{a \in A \mid M, a \models \textbf{all}(\{y \mid \mathrm{R}(y,x)\}\,\{z \mid \mathrm{S}(z)\}))\} =$$
$$\{a \in A \mid ([\![\{y \mid \mathrm{R}(y,x)\}]\!]_{M,a}, [\![\{z \mid \mathrm{S}(z)\}]\!]_M) \in M_Q(\textbf{all})\} =$$
$$\{a \in A \mid [\![\{y \mid \mathrm{R}(y,x)\}]\!]_{M,a} \subseteq [\![\{z \mid \mathrm{S}(z)\}]\!]_M\} =$$
$$\{a \in A \mid \{b \in A \mid M, b, a \models \mathrm{R}(y,x)\} \subseteq \{c \in A \mid M, c \models \mathrm{S}(z)\}\} =$$
$$\{a \in A\{(b,a) \mid (b,a) \in M_R(R)\} \subseteq \{c \in A \mid (c) \in M_R(S)\}\}$$

Suppose that $M_R(R)$ is empty. The problem is that in such a situation,

$$\{b \in A \mid (b,a) \in M_R(R)\} = \emptyset$$

for any value $a \in A$. But given the semantics of **all**, if the first set is empty, the quantifier is satisfied, regardless of what the second argument is. Therefore, whenever $\mathcal{U}$ is infinite, the query would have an infinite answer, since any value $a \in A$ will qualify. But even if $M_R(R)$ is not empty, consider any value $a \in A$ such that there is no $b \in A$ with $(b, a) \in M_R(R)$. Thus, the set term associated with a will be empty. But, once again, this means that a actually qualifies for the meaning of the GQ **all**. Note that this is not a problem of **all** only, but it happens any time a parametric set is an argument for a downward monotonic GQ[7]. But since whenever $a \notin adom(M)$ we have that there will be no b with $(b, a) \in M_R(R)$, this implies that the language, as defined, is also *not* domain independent.

Example 4.8. Consider a query over the same vocabulary:

$$\{x \mid \textbf{no}$$
$$(\{z \mid \mathrm{R}_2(z) \wedge z = x\})\}$$

*In any given γ_2-interpretation where the extension of relations is finite and the universe is infinite, the above query is unsafe. Observe that by looking at at the formula $\mathrm{R}_2(z) \wedge z = x$ one cannot tell that the formula is unsafe. Such a formula would be considered safe on many approaches, including [98]. The problem lies in the semantics of the quantifier **no**, which make the formula true only when the set term is empty. Any value in the universe not in the extension of R qualifies for the answer.*

---

[7] Remember that a GQ Q is downward monotone on the i-th argument if $Q(A_1, \ldots, A_i, \ldots, A_n)$ and $B \subseteq A_i$ imply $Q(A_1, \ldots, A_{i-1}, B, A_{i+1}, \ldots, A_n)$. Note that this implies that the empty set is an argument for the quantifier.

There are basically two kinds of approaches to these problems: a *semantic approach* and a *syntactic approach*. The semantic approach consists on changing the domain of evaluation.

Definition 4.9. $QLGQ^{adom} = \{[\![\varphi]\!]_{adom}| \varphi \in QLGQ\}$ is the language $QLGQ$ under active domain semantics.

Lemma 4.10. *Let $\mathcal{Q}$ be an arbitrary set of GQs. Then $QLGQ^{adom}(\mathcal{Q})$ is safe.*

This is a trivial result since, for any interpretation D, $adom(D)$ is a finite set and therefore the semantics of any query will yield a finite set. It is also an unsatisfactory approach, as it does not tell us anything about which parts of the language contribute to the problem and how. The syntactic approach is more complicated, but it is more constructive, so we follow it next.

Definition 4.11. Let $\gamma = [(\mathcal{R}, \alpha), \mathcal{C}, (\mathcal{Q}, \beta)]$ be a *GQ*-vocabulary. The language $LQLGQ(\gamma)$ is defined as follows:

1. *Basic terms*: Basic terms are defined as in $QLGQ$.
2. *Set terms*: Set terms are defined as in $QLGQ$.
3. *Formulas*
    a) If ψ is a formula, and t_1, t_2 are terms, then $(\psi \wedge t_1 \; \theta \; t_2)$ is a formula, provided that
        - θ is equality, and $Fvar(t_1 = t_2) \cap Fvar(\varphi) \neq \emptyset$, or
        - $Fvar(t_1 \neq t_2) \subseteq Fvar(\psi)$.
    b) If φ_1 and φ_2 are formulas then $\varphi_1 \wedge \varphi_2$ is a formula.
    c) If $R \in \mathcal{R}$, $\alpha(R) = n$ and $t_1, \ldots, t_n$ are basic terms, then $R(t_1, \ldots . t_n)$ is a basic formula.
    d) If $Q \in \mathcal{Q}$ is a quantifier name of type $[k, m]$ (i.e., $\beta(Q) = [k, m]$) and $S_1, \ldots, S_k$ are set terms of arity m, and ψ is a formula, then $Q(S_1, \ldots, S_k) \wedge \psi$ is a basic formula, provided that $Fvar(S_1) \cup \ldots \cup Fvar(S_k) = Fvar(\psi)$.

We also define $LQLGQ(\gamma)$ sentences and queries.

Definition 4.12. • A $LQLGQ(\gamma)$ *sentence* τ is a formula of $LQLGQ(\gamma)$ without free variables (i.e. $Fvar(\tau) = \emptyset$).
- A $LQLGQ(\gamma)$ *query* is a set term S of $LQLGQ(\gamma)$ without free variables (i.e. $Fvar(S) = \emptyset$).

We call this language $LQLGQ$ for *Limited QLGQ*. Note that the syntax of $LQLGQ$ is more restrictive than that of $QLGQ$; that is, all formulas of $LQLGQ$ are also formulas of $QLGQ$. This allows us to define the semantics of $LQLGQ$ as that of $QLGQ$ applied to the valid formulas of $LQLGQ$.

Theorem 4.13. *em For all vocabularies γ, all γ-interpretations $M = [A, M_r, M_Q]$, all queries $q \in LQLGQ(\gamma)$ with arity n, $[\![q]\!]_M \subseteq adom(M)^n$.*

Corollary 4.14. *$LQLGQ$ is a safe language.*

As explained above, domain independence is a highly desirable property on query languages, and we would like to get a language that possesses it. However, in example 4.6 a query in $QLGQ(\mathbf{some^1})$ was introduced that is not domain independent. The culprit is the $\mathbf{all^1}$ quantifier. When changing the domain, the *global* quantifier remains the same, but the *local* quantifier changes with the domain. Nothing in the definitions guarantees that $Q(S)$, the extension that quantifier Q has in S, is consistent with $Q(A)$, the extension of Q in $\mathcal{U}$. Intuitively, one expects Q to have the same *intension* on all domains. To capture this idea, we restrict our attention, for the following lemma, to quantifiers that are *domain independent*:

Definition 4.15. A quantifier Q of type $[k_1, \ldots, k_n]$ is domain independent if for any $M \subseteq M' \subseteq \mathcal{U}$, $Q_M = Q_{M'} \uparrow M$.

As both $Q(M)$ and $Q(M')$ are sets of sets, recall from the conventional notations that $Q(M') \uparrow M$ must be understood as $\{(A_1, \ldots, A_n) \in Q(M') \mid A_i \subseteq M^{k_i}, 1 \leq i \leq n\}$.

Example 4.16. The quantifier $\mathbf{all^1}$, defined, for all $M \subseteq \mathcal{U}$, by $\mathbf{all^1}(M) = \{A \subseteq M \mid A = M\} = \{M\}$, is not domain independent, as for any $M' \subset M$, $\mathbf{all^1}(M) \cap \{M'\} = \emptyset$.

It is not difficult to see that this property is equivalent to what we called EXT in chapter 3.

From now on, we will restrict our attention to domain independent quantifiers. With this proviso we have that, for any M such that $adom(D) \subseteq M \subseteq \mathcal{U}$, the γ-interpretation generated by M $D_M = (M, D_{\mathcal{R}}, D_{\mathcal{Q}} \uparrow M)$ is such that $D_{\mathcal{Q}} \uparrow M$ can be defined as $\{Q(D) \uparrow M \mid Q \in D_{\mathcal{Q}}\}$, as before, or a $\{Q(M) \mid Q \in D_{\mathcal{Q}}\}$; both formulations are equivalent for quantifiers that respect ISOM and EXT. An important consequence of this is the following:

Proposition 4.3.1 Let D be an interpretation and let M be such that $adom(D) \subseteq M \subseteq \mathcal{U}$ and M is infinite enumerable. Then D_M, the interpretation originated by M, is isomorphic to D.

We can now finally state our main result:

Theorem 4.17. *Let $\mathcal{Q}$ be any set of quantifiers that follow ISOM and EXT. Then $LQLGQ(\mathcal{Q})$ is domain independent.*

Even though the semantic and syntactic approach appears completely different, it turns out that they are equivalent:

Theorem 4.18. *Let $\mathcal{Q}$ be any set of quantifiers that follow ISOM and EXT. Then $LQLGQ(\mathcal{Q}) \equiv QLGQ(\mathcal{Q})^{adom}$.*

However, the syntactic approach, as argued above, is more constructive in that it allows us to define a safe, domain independent language that works in finite and infinite interpretations. In the rest of this book, we use the language $LQLGQ$ and restrict our quantifiers to those that follow ISOM and EXT.

4.3.1 Relation to other languages

It is natural to ask whether other logic-based query languages behave in a similar manner with respect to safety and domain independence.

Following Hull and Su ([53]), we define, given interpretation $M = (A, R_1, \ldots, R_n)$, query language $\mathcal{L}$, $\|\bar{\varphi}\|_{oru}$, for any formula $\varphi \in \mathcal{L}$, as $[\![\bar{\varphi}]\!]_A \cap adom(M)$; $[\![\bar{\varphi}]\!]_{finv}$ as $[\![\varphi]\!]_S \cap adom(M)$, where S is finite and $adom(M) \subseteq S \subseteq \mathcal{U}$; and $[\![\varphi]\!]_{cinv}$ as $[\![\varphi]\!]_S \cap adom(M)$, where S is infinite enumerable and $adom(M) \subseteq S \subseteq \mathcal{U}$. Correspondingly, the language $\mathcal{L}^{oru}$ is the set of queries expressible by formulas of $\mathcal{L}$ under the corresponding semantics; that is, $\{[\![\varphi]\!]_{oru} \mid \varphi \in \mathcal{L}\}$. The definition is extended to $\mathcal{L}^{finv}$ and $\mathcal{L}^{cinv}$. We define $\mathcal{L}^{ind}$ as the set of formulas in $\mathcal{L}$ that are domain independent[8].

Theorem 4.19. *([53]) Let CALC be the relational domain calculus. Then*

$$CALC^{adom} \equiv CALC^{oru} \equiv CALC^{cinv} \equiv CALC^{finv} \equiv CALC^{ind}$$

Given the results above, we pose the question of whether this result can be extended to our language.

Question Is it the case, for sufficiently general set of quantifiers $\mathcal{Q}$, that $QLGQ(\mathcal{Q})^{adom} \equiv QLGQ(\mathcal{Q})^{oru} \equiv QLGQ(\mathcal{Q})^{cinv} \equiv QLGQ(\mathcal{Q})^{finv} \equiv QLGQ(\mathcal{Q})^{ind}$?

A partial answer to the question follows.

Corollary 4.20. *Let $\mathcal{Q}$ be a set of quantifiers such that all its elements follow ISOM and EXT. Then $LQLGQ(\mathcal{Q})^A \leq QLGQ(\mathcal{Q})^{ind}$.*

This is a corollary of Theorem 4.17.

Corollary 4.21. *Let $\mathcal{Q}$ be a set of quantifiers such that all its elements follow ISOM and EXT. Then $QLGQ(\mathcal{Q})^{ind} \equiv QLGQ(\mathcal{Q})^{adom}$.*

Proposition 4.22. *Let $\mathcal{Q}$ be a set of quantifiers such that all its elements respect EXT, and assume our domain of discourse A is countable. Then $QLGQ(\mathcal{Q})^{cinv} \equiv QLGQ(\mathcal{Q})^{oru}$.*

Proposition 4.23. *Let $\mathcal{Q}$ be a set of quantifiers such that all its elements respect EXT. Then $QLGQ(\mathcal{Q})^{ind} \leq QLGQ(\mathcal{Q})^{oru}$.*

It is unknown at this time whether $QLGQ(\mathcal{Q})^{xxx} \leq QLGQ(\mathcal{Q})^{yyy}$, where xxx stands for oru or $cinv$ and yyy stands for ind or $adom$. It is even unknown whether $QLGQ(\mathcal{Q})^{cinv} \equiv QLGQ(\mathcal{Q})^{finv}$. Hull and Su establish these results for CALC with a tricky construction that simulates the semantics with invented values in the active domain. For that, a complicated transformation that examines the semantics of $\forall$ and $\exists$ is necessary. In the present context, the class $\mathcal{Q}$ of quantifiers that respect ISOM and EXT is too unruly to allow such a construction. In any case, and given the high expressive power of such a class, we conjecture a negative answer to the question posed.

---

[8] [53] use $\mathcal{L}^{lim}$ to denote $\mathcal{L}^{adom}$ (the language under active domain semantics).

4.4 Generalized Quantifiers and SQL

From our viewpoint, the main weakness of SQL is due to its relational algebra roots: it focuses on *tuple-to-tuple* comparisons, that is, it only allows comparison of values in one particular tuple to another particular tuple. The tuples are first aligned with a join, and then a selection condition is applied to the resulting tuple. This can be extended without problem to a *given* number of tuples, that is, given a number n of tuples, we can write an SQL query that will compare the values in n tuples. However, it is much trickier to deal with queries where an arbitrary *set* of tuples must be considered. SQL offers *nested* or *embedded* subqueries, which offer *tuple-to-set* comparisons, using predicates like IN, NOT IN, or comparisons with SOME (ANY) or ALL. However, this extended ability is simply syntactic sugar; techniques for *unnesting* have shown that any SQL query with nested subqueries can be written as an SQL query without subqueries[9]. Thus, while making some queries simpler to write (a positive development, no doubt), this extension still falls short of what is needed in certain cases. We contend that SQL needs to take one further step and allow the *direct* expression of *set-to-set* comparisons. As we will see, it will turn out that this extension is also syntactic sugar for SQL (but not for relational algebra!). A previous extension to SQL, the ability to aggregate (in particular, to count) and to group, gives us the ability to *indirectly* capture most relations between sets (not between arbitrary relations) -we will make this assertions precise in chapter 5.4. Nevertheless, we contend that this is still an improvement for the language, as it not only makes queries easier to write, but also easier to optimize. And it may be a stepping stone for more ambitious extensions, like a (limited) amount of nesting. Thus, taking into account that added expressive power brings with it higher complexity, one cannot simply throw in new, powerful operators to a query language; rather, particular operators must be chosen to give the adequate balance between power and complexity[10].

Hsu and Parker ([52]) have proposed the incorporation of generalized quantifiers into SQL. After reviewing some of the problems with SQL's treatment of quantification, they introduce the concept of Generalized Quantifier as binary relations between sets and show how its use greatly simplifies the expression of queries involving quantification. Their proposal is to introduce *GQ*s in the WHERE clause of SQL queries, where the usual syntax for subqueries is augmented, in order to specify relations between sets with *GQ*s. The sets are specified with regular SQL subquery. After that, a preprocessor transforms the query with *GQ*s in one without. The resulting query can then be processed by a standard relational database system.

The approach of Hsu and Parker has some obvious advantages. First, by making the quantifiers to be, basically, macros on top of standard SQL, they

---

[9] There are some technical subtleties to be taken care of, like the issues with duplicates or the *zero-sum* bug, but these are well known and can be dealt with.

[10] See subsection 9.1.3 for an extension that throws off the balance.

offer a very easy way to implement the GQs on top of current relational systems. Second, their approach does allow to express complicated queries in a more elegant, simpler manner when compared to standard SQL. Third, the queries in the new language are automatically safe, because SQL is safe and the extended syntax does not allow the *GQ*s to be used to create any extra values. For instance, our first and second example queries become, in Hsu and Parker's extension,

```
SELECT pid FROM Professor P
WHERE ALL
      (SELECT sid FROM Student)
      (SELECT sid FROM Teaches WHERE pid = P.pid)
```

and

```
SELECT pid FROM Professor P
WHERE NO
      (SELECT sid FROM Student)
      (SELECT sid FROM Teaches WHERE pid = P.pid)
```

Observe, in particular, that in both queries a change in a single determiner provokes only a change in the quantifier. We note the similarity between these queries and $QLGQ$ queries: the set terms in $QLGQ$ can be seen as basic query blocks in SQL -and therefore as subqueries also. In particular, parametric set terms correspond to *correlated* subqueries. In $QLGQ$, though, the use of logical notation does not correspond to the naming notation used by SQL. This brings one fundamental difference in that, in SQL, we need a separate query block simply to introduce the correlation variables to be used later. Thus, there is no counterpart in $QLGQ$ to the first two lines of the SQL query: SELECT pid FROM Professor... since their main purpose is, as stated, to create the environment where the correlated variable pid in the second subquery can be obtained. Note that this forces SQL to introduce a join (or outerjoin) between the table in the outer query and the subquery (in this example, a join between Professor and Teaches). For SQL to recognize this as a join, unnesting techniques may be used -a naive evaluation would be extremely costly. Note also that when the join is, as in this case, between a primary key in the outer block and a foreign key in the inner (nested) block, it can be totally avoided. But this join cannot be avoided without deep changes to the SQL syntax. On the other hand, if one was interested in only certain values of the correlated variable, in $QLGQ$ such restrictions would be introduced within the (parametric) set terms, while in SQL they could be introduced in the main (outer) query or in the subquery. It is, of course, much more advantageous to use such restrictions to constrain the set of values considered for correlation, and that is exactly what the *magic set* approach does ([78, 79]). To see the difference, assume the relation Professor has an attribute rank, and that in the query above we are only interested in full professors. Then we could write

```
SELECT pid FROM Professor P
WHERE  P.rank = ''full'' and
       ALL
       (SELECT sid FROM Student)
       (SELECT sid FROM Teaches WHERE pid = T.pid)
```

instead of the query above. Then the magic set technique would restrict the set of professor ids considered -effectively pushing down a selection before the join created by unnesting. In *QLGQ*, however, this query would have to be written as

$$\{y \mid \mathbf{all}\{x \mid \text{Student}(x)\}, \{x \mid \text{Teaches}(y, x) \wedge Professor(y,' full')\}\}$$

Note that here the selection is naturally "pushed" since it is expressed within the parametric set term. Hence, the effect of the magic approach is achieved naturally without any need for further optimization.

Thus, one could argue that the improvements of the approach in [52] are somewhat limited limited as the *GQ*s still have to interact with SQL's (limited) ability to define and manipulate sets. Since SQL is tuple oriented, forming a set must be done using a variable to name the elements in the set. This forces SQL to use complicated syntax (subqueries) to form certain set expressions.

However, this problem pales when compared to two other, more serious issues. First, when translated back into SQL, the queries become quite complex again, and cannot be executed efficiently by commercial relational engines (the authors acknowledge that "we cannot claim that a translated extended SQL query is faster than a query written in SQL2" -see [52], section 4). However, one could argue (and rightly so, in our opinion) that now the work has been displaced to that of finding good optimization techniques for this kind of query, and the the goal at hand (to simplify the queries for the end user) has been accomplished. [52] points out some possible optimizations, but those are limited to avoiding repeated subexpressions. In [51], some examples of further possible optimizations are shown, based on the laws regulating equivalence of expressions with *GQ*s can be derived from the properties of *GQ*s seen as binary relations. However, it is not shown how these optimizations would be integrated in a regular relational optimizer. Thus, the issue of efficient implementation and optimization need to be considered in more depth.

A second (and more fundamental) drawback of this approach is that both the type and kind of quantifiers considered is limited. The idea of *GQ* used in [52] and in [51] has no concept of *type*; only standard (type $(1, 1)$) quantifiers are used. Also, only quantifiers following CONS (and implicitly, ISOM) are considered. Finally, quantifiers are defined directly, by giving a list. Thus, only a fixed set of quantifiers is offered. Therefore, Hsu and Parker's approach does not use the full potential and flexibility of the notion of Generalized Quantifier. The author suggests that "... a better approach appears to be allowing the set of quantifiers to grow as necessary for particular situations, applications, or user. For example, the grammar could be made extensible with a dynamically modified *macro* facility..." [52][page 105]. But there is no general framework in [52] in which to carry out this task. To provide one such framework has been one of the goals of this work. In the next chapter we will see how to define *collections* of quantifiers which can then be added to the language. We will also see that all quantifiers in the collection can be efficiently implemented. To achieve this, we will use some basic properties of Generalized Quantifiers that logicians have discovered ([104, 65]), and exploit the generality of their definitions. However, the matter of how to use such collections in SQL is still a pragmatic issue. While we pointed out the inadequacies of past approaches, one must realize the deep changes to a standard are difficult -and unlikely. Thus, one must still work with a restricted form of generalized quantification that allows their use in a manner similar to the one seen above.

5

Implementation and Optimization of Standard GQs

5.1 Languages to Define GQs

For the rest of this chapter, we will restrict our attention to standard quantifiers; that is, monadic binary quantifiers (of type $(1,1)$). In addition, we will make one further restrictions on the quantifiers to be considered in this section: we will assume that they obey EXT. This was defined in chapter 3; here we repeat definition 3.2.5 simplified to the $(1,1)$ type:

Definition 5.1.1 (EXT) A quantifier Q of type $(1,1)$ follows EXT if for all M, M', $A_1, A_2 \subseteq M \subseteq M'$, $Q_M(A_1, A_2)$ iff $Q_{M'}(A_1, A_2)$.

What will allow us to define effective ways to deal with these quantifiers is the following: when dealing with monadic quantifiers, all arguments are sets. Sets do not carry structural information, like relations do. Thus, isomorphisms between sets are simply bijections. In other words, all that matters to check if sets are isomorphic is the *cardinality* of the sets. When dealing with finite models, as we do, all sets are finite. Thus, we restrict ourselves to *finite quantifiers*, those where all arguments are finite sets. Hence, matters of cardinality can be seen as simply arithmetic predicates on natural numbers. To define quantifiers, then, we can give properties on natural numbers.

To put this idea formally, recall that a (local) quantifier Q of type $(1,1)$ can be identified with a class of structures $\mathcal{M} = < M, A_1, A_2 >$, where M is our universe of discourse and $A_1, A_2 \subseteq M$. Since Q is invariant under isomorphism, any structure isomorphic to $\mathcal{M}$ is also an element of the class.

We define, for any $A \subseteq M$, $A^1 = A$ and $A^0 = M - A$. Let s be a function from $\{A_1, A_2\}$ into $\{0, 1\}$, and let $\mathcal{S}$ be the set of all such functions. Define $P_s^M = A_1^{s(A_1)} \cap A_2^{s(A_2)}$ for any $s \in \mathcal{S}$. Then the set $\{P_s^M\}_{s \in \mathcal{S}}$ is a *partition* of M. This partition is all that can be said about $\mathcal{M}$ from a logical point of view; that is, for any two monadic structures $\mathcal{M}$, $\mathcal{M}'$ (both of type $(1,1)$), for all $s \in \mathcal{S}$, if $|P_s^M| = |P_s^{M'}|$, then $\mathcal{M}$ and $\mathcal{M}'$ are isomorphic ([102]). Since Generalized Quantifiers, as classes of structures, are closed under isomorphism, we have the following:

A.Badia, *Quantifiers in Action: Generalized Quantification in Query. Logical and Natural Languages*, Advances in Database Systems 37, DOI: 10.1007/978-0-387-09564-6_5,
© Springer Science+Business Media, LLC 2009

Proposition 5.1.2 *If Q is a standard GQ and $\mathcal{M} = \langle M, A_1, A_2 \rangle$ and $\mathcal{M}' = \langle M', A_1', A_2' \rangle$ are such that for all $s \in \mathcal{S}$, $|P_s^M| = |P_s^{M'}|$, then $Q_M(A_1, A_2)$ iff $Q_{M'}(A_1', A_2')$.*

For structures of type $(1, 1)$, the elements of the partition defined above are $A_1 \cap A_2$, $A_1 - A_2$, $A_2 - A_1$, and $M - (A_1 \cup A_2)$. Thus, any two structures (quantifiers) $\langle M, A_1, A_2 \rangle$ and $\langle M', A_1', A_2' \rangle$ are isomorphic (and therefore define the same quantifier) iff these four elements have the same cardinality. Hence, the above can be rephrased as follows:

Lemma 5.1. *If Q is a standard GQ, and $\mathcal{M} = \langle M, A_1, A_2 \rangle$, $\mathcal{M}' = \langle M', A_1', A_2' \rangle$ are such that $|M - (A_1 \cup A_2)| = |M' - (A_1' \cup A_2')|$, $|A_1 - A_2| = |A_1' - A_2'|$, $|A_2 - A_1| = |A_2' - A_1'|$ and $|A_1 \cap A_2| = |A_1' \cap A_2'|$, then $Q_M(A_1, A_2)$ iff $Q_{M'}(A_1', A_2')$.*

This can be simplified for quantifiers that follows EXT:

Lemma 5.2. *If Q is a standard GQ that respects EXT, and $\mathcal{M} = \langle M, A_1, A_2 \rangle$, $\mathcal{M}' = \langle M', A_1', A_2' \rangle$ are such that $|A_1 - A_2| = |A_1' - A_2'|$, $|A_2 - A_1| = |A_2' - A_1'|$ and $|A_1 \cap A_2| = |A_1' \cap A_2'|$, then $Q_M(A_1, A_2)$ iff $Q_{M'}(A_1', A_2')$.*

Intuitively, EXT allows us to ignore the context (anything outside) of A_1 and A_2.

Remark The definition above can be easily extended to any monadic quantifier, regardless of the number of arguments. Focusing on the $(1, 1)$ type, one may be surprised that a structure $\langle M, A_1, A_2 \rangle$ does not need to be represented by all Boolean combinations of their elements. Intuitively, one could look at the Venn diagram of such an structure and realize that any closed region on the diagram can be denoted by a formula in any language at least as powerful as FOL -for instance, $A_1 \cup A_2$ can be captured by a disjunction. There are 2^4 such regions for this structure: first, we have $\emptyset$, A_1, A_2, $A_1 \cap A_2$, $A_1 \cup A_2$, $A_1 - A_2$, $A_2 - A_1$, $(A_1 - A_2) \cup (A_2 - A_1)$. In addition, for any such region B, there is $M - B$ (note that $M - \emptyset = M$). However, the 4 regions that we use are enough to capture all the relevant information. It can be easily seen that

- $A_1 = (A_1 - A_2) \cup (A_1 \cap A_2)$, and similarly for A_2.
- $A_1 \cup A_2 = (A_1 - A_1) \cup (A_2 - A_1)$
- $M = (M - (A_1 \cup A_2)) \cup (A_1 - A_2) \cup (A_2 - A_1)$; and each set of the form $M - B$ can be recovered similarly.

Since the regions used are disjoint, it is also easy to see that the respective cardinality relations also hold; for instance, for the first equivalence above, we have that $|A_1| = |(A_1 - A_2)| + |(A_1 \cap A_2)|$.

Q.E.D.

From the above lemmas it follows that any quantifier of type $(1, 1)$ is determined by a relation between the cardinalities of the elements of the partition, in the following sense (this concept is called *number-theoretic definability* in [102]).

Definition 5.1.3 Let Q be a standard quantifier, and A_1 and A_2 sets. Let $c_1^{A_1,A_2} = |A_1-A_2|$; $c_2^{A_1,A_2} = |A_2-A_1|$ and $c_3^{A_1,A_2} = |A_1\cap A_2|$, and let $\varphi(x,y,z)$ be a formula in some arithmetic language with (at least) 3 free variables. We say that Q is *number-definable* if

$$Q(A_1,A_2) \; iff \; \varphi(c_1^{A_1,A_2}, c_2^{A_1,A_2}, c_3^{A_1,A_2})$$

We now have a method to define collections of quantifiers: by giving a particular language on numbers. Then, all quantifiers definable by formulas in the given language are in the collection. In this section, we give three closely related number languages.

We begin with a pair of logics proposed by Westerståhl ([102]):

Definition 5.1.4 *([102])* The *pure number language* (*PNL*) is defined as follows:

1. Terms and variables.
    a) $0,1,2,\ldots$ are numerals.
    b) $x,y,z,\ldots$ are natural number variables.
2. Formulas.
    a) If n is a numeral, and x is a natural number variable, $x = n$ is a formula.
    b) If φ and ψ are formulas, then $\neg\varphi$, $\neg\psi$, $\varphi\wedge\psi$, $\varphi\vee\psi$ are formulas.

Definition 5.1.5 *([102])* The *proportional language* (*PL*) is the set of formulas on natural numbers obtained under the following syntax:

- Terms and variables
    - $0,1,2,\ldots$ are numerals.
    - If m,n are numerals, $m+n$ is a numeral.
    - If m,n are numerals, $m\times n$ is a numeral
    - $x,y,z,\ldots$ are domain variables.
    - Numerals and variables are terms.
- Formulas
    - If t_1 and t_2 are terms, $t_1=t_2$ is a (atomic) formula.
    - If φ, ψ are formulas, then $\neg\varphi$, $\neg\psi$, $\varphi\wedge\psi$, $\varphi\vee\psi$ are formulas.

Nothing else is a formula[1].

---

[1] A very similar family of quantifiers is called the *properly proportional* quantifiers in [61]. A *properly proportional* quantifier $Q_{m,n}^{\Theta}(A,B)$ is defined by an equation of the form

$$\frac{|A\cap B|}{|B|} \; \Theta \; \frac{m}{n}$$

where m,n are natural numbers, $n\neq 0$ and Θ is one of $=,\neq,\leq,\geq$. Since $x_3 = |A\cap B|$ and $x_3+x_6 = |B|$, we can write the equation as $x_3\times n \; \Theta \; (x_3+x_6)\times m$.

Definition 5.1.6 The *existential proportional language (EPL)* is the set of formulas on the natural numbers obtained under the following syntax:

- If φ is a formula of PL, then φ is a formula of EPL.
- If φ is a formula of EPL *where the symbol* $\times$ *does not appear*, and x is a variable, $\exists x \varphi(x)$ is a formula of EPL.

The traditional descriptions of free and bound variables apply. Nothing else is a formula.

Clearly, each language is a bit more powerful than the previous one(s). Some examples of GQs definable in PNL are:

- **all**(A_1, A_2), defined by $x_1 = 0$;
- **some**(A_1, A_2), defined by $\neg(x_3 = 0)$;
- **exactly 2**(A_1, A_2), defined by $x_3 = 2$;
- **at most** $k(A_1, A_2)$, defined by $x_3 = \mathbf{1} \vee \ldots \vee x_3 = \mathbf{k}$.

However, an analysis of PNL reveals that it is a weak language that cannot express many relations between numbers. PL, on the other hand, is quite a powerful language. PL can express *exactly half the A_1s are A_2s* (with the formula $x_3 = x_5$), and the Hartig quantifier (with the formula $x_5 = x_6{}^2$). Neither quantifier is expressible in PNL. Also, PL can write formulas like the following:

$$\frac{n}{m} \text{ of } (A_1, A_2) = x_3 \times m = x_1 \times n$$

with the meaning $\frac{n}{m}$ *of A_1s are A_2s*. When $m = 100$, we have the *percent* family of quantifiers. This also allows PL to express *proportional GQs*, like **one half of, one third of...**.

Finally, EPL can express the quantifier **more**(A_1, A_2) (defined as $(|A_1| > |A_2|)$) as

$$\exists n \ x_1 = x_2 + n \wedge \neg(n = \mathbf{0})$$

and the **even** quantifier (since A_1 *is even* can be expressed as $\exists n \ x_1 = n + n$)[3] and is in general more powerful than PL[4]. Still, EPL has been restricted to not allow quantification over multiplication in order to keep it decidable.

Remark A question arises as to what logic do these language capture, that is, if we allow basic arithmetic on the signature of a logic language, which logic languages correspond to the number languages give above? The following formalizes the question:

---

[2] Note that $|A - B| = |B - A|$ implies $|A| = |B|$.

[3] From some of the examples it is obvious that these languages do not have a 0-1 law.

[4] Note that $m \leq n$ can be written in EPL as $\exists z \ m + z = n$.

Definition 5.1.7 Q is *logic definable* in a given logic $\mathcal{L}$ if there exists a sentence ψ_Q in $\mathcal{L}$ such that for all finite structures $< M, A_1, A_2 >$,

$$Q_M(A_1, A_2) \Leftrightarrow < M, A_1, A_2 > \models \psi_Q$$

Let *FOC* (*First Order Logic with Counters*) be the language defined by extending *FOL* with variables and constants for the natural numbers, interpreted over a two-sorted structure (one where there is a set isomorphic to $\mathbb{N}$, the natural numbers, besides the traditional universe). Formulas involving number variables and constant are called *counting terms*. We say that $FOC^=$ is the language where the only relations between counting terms is equality (in other words, no arithmetic is allowed). Moreover, the only formulas allowed are the usual *FOL* formulas formed by conjunction, disjunction, negation and existential quantification of free variables (number variables or otherwise). We obtain the language $FOC^{=,\leq}$ by adding formulas of the type $C_1 \leq C_2$, where C_1, C_2 are counting terms, to $FOC^=$. Similar definitions lead to $FOC^{=,\leq,+}$ and $FOC^{=,\leq,+,\times}$. In this last case we have the important proviso that quantification is not allowed in formulas that contain $\times$; this allows us to keep the language decidable, since we basically have Presburger arithmetic[5].

Over monadic structures, we obtain the following:

Theorem 5.1.8 ([102]) Let Q be a standard quantifier. Then the following statement are equivalent:

1. Q is logic definable in $FOC^=$.
2. Q is number definable in PNL.

A proof of this result can be found along the lines of [102]. Westerståhl proves his result for a more restrictive class of GQs, one that obeys ISOM, EXT and CONS. As a consequence, his numerical definition needs only two arguments. Even though his definition is more restrictive, close examination of the proof shows that it still holds in the present context.

This explains why PNL is such a restricted language; it basically captures only the counting ability of *FOL*, which is very limited. We can see how some quantifiers defined above in PNL are definable in $FOC^=$:

- **all**(A_1, A_2), defined by $\neg \exists x(\neg A_1(x) \wedge A_2(x))$;
- **some**(A_1, A_2), defined by $\exists x(A_1(x) \wedge A_2(x))$;
- **exactly 2**(A_1, A_2), defined by $\exists x \wedge \exists y(x \neq y \wedge A_1(x) \wedge A_2(x) \wedge A_1(y) \wedge A_2(y))$; and
- **at most** $k(A_1, A_2)$, defined by $\exists x_1 \ldots \exists x_k(\bigwedge_{i=1}^{k}(A_1(x_i) \wedge A_2(x_i)) \wedge \neg \exists z(\bigwedge_{i=1}^{k}(z \neq x_i)))$.

Theorem 5.1.9 Let Q be a standard quantifier. Then the following statement are equivalent:

---

[5] These languages are basically defined in [40].

1. Q is logic definable in $FOC^{=,\leq,+,\times}$.
2. Q is number definable in EPL.

   Q.E.D.

From now on, we assume that any generalized quantifier that one wants to make available for $QLGQ$ is number definable, and that a number formula is provided to define the quantifier. The challenge now is to take arbitrary number formulas in a given language and use this information to generate queries over a relational database, that is, in relational algebra. This is the task that we turn to next.

5.2 Translating and Optimizing $QLGQ$

As stated when the language was introduced, $QLGQ$ is really a parametric language. The actual language is given by introducing a particular set of GQs to be used. One good choice is that of the type (1,1) quantifiers definable in EPL. This is a reasonably powerful class, while admitting an efficient implementation.

Our strategy is to define an *interpreter* that takes as input a particular $QLGQ$ query φ and has access to (at least) one definition in PL for each quantifier used in φ. The interpreter then produces as output an expression in an extended version of the relational algebra. This extended version is simply the traditional relational algebra plus

- a *group by* operator (denoted GB), with the same semantics as the GROUP BY clause in SQL;
- the count aggregate function (again, with the same semantics as SQL); and
- the CASE operator from the new SQL standard.

Clearly, the goal of our interpreter is to generate an expression that can easily be translated into SQL. Why not use SQL directly as our target language? Unfortunately, SQL lacks the orthogonality of the relational algebra, and it forces the user to break down complex expressions that can be expressed as a single "block" in relational algebra. Thus, our choice of target language simplifies not only the task of generating interpretations for $QLGQ$ formulas, but also (more importantly) of applying optimizations to them.

5.3 The Interpreter

From now on, we assume that for any quantifier Q appearing in a $QLGQ$ query φ, the system has a PL formula that defines Q. Note that there may be more than one formula for a given quantifier; this fact can be used for optimization purposes as follows: each definition of the quantifier can be used to generate a translation of the formula using the quantifier. Because each translation is an expression in extended relational algebra, we will be able to estimate a cost for executing it. As in standard cost-based optimization, we can then choose the translation with the smallest cost among all available.

Our target language is an extended relational algebra, with aggregation, group by and the ability to detect nulls (an is null predicate) and to count under certain simple conditions (that is, a CASE statement). This algebra is similar to that used by most query processors and optimizers, a fact that will be used later.

We now show how to generate an extended relational algebra expression for any $QLGQ$ query φ. If φ does not involve GQs, then it is of the form $\{\overline{x} \mid \psi(\overline{x})\}$, where ψ is a simple formula, or a conjunction of simple formulas. Looking at the semantics of $QLGQ$, it is clear that such queries correspond to SPJ queries and that a translation for them can be obtained easily. The interesting case involves dealing with GQs; therefore, we focus our attention on the case where ψ is of the form $Q(X_1, X_2)$, with Q a GQ and X_1, X_2 set terms. We will apply our interpreter recursively, from the bottom up, and hence assume that we have already obtained two relational algebra expressions E_1 and E_2 that correspond to X_1 and X_2, respectively. The interpreter then simply uses the PL formula associated with Q (any one of them, if several are available), and proceeds as follows: first, generate the expression $E_1 - E_2$ if c_1 is involved; $E_2 - E_1$ if c_2 is involved; and $E_1 \cap E_2$ if c_3 is involved (note: we can use join instead of intersection, and will do so in the following). Then, using aggregation and count over the expression generated to obtain a corresponding counter, a selection is added with a condition that reflects the PL formula. As an example, assume the query
$$\{x \mid \textbf{at least 2}(\{y \mid \varphi(y, x)\}, \{z \mid \psi(z)\})\}$$
that would produce the following formula:
$$\{\pi_A(\sigma_{c \geq 2}(GB_{A, count(distinct\ B)\ as\ c}(R \bowtie_{B=D} S)))\}$$
where $R(A, B)$ is the result of evaluating $\{y \mid \varphi(y, x)\}$, and $S(D)$ is the result of evaluating $\{z \mid \psi(z)\}$. Since R corresponds to a parametric set, we divide R into groups using the values of the parameter (A) (that is, we group by A). Note that the use of *distinct* with our count, motivated by the fact that GQs are predicates on *sets*[6].

While it is not difficult to relate the translation to the original semantics, it turns out that there are several subtleties that prevent us from generalizing from this example. First, there is a *zero sum bug* [34] latent in $QLGQ$ as in SQL: when a quantifier condition involves zero, problems may arise. Assume, for instance, quantifier **no**, that can be defined by $c_3 = 0$, and query
$$\{x \mid \textbf{no}(\{y \mid \varphi(y, x)\}, \{z \mid \psi(z)\})\}$$
The previous approach would generate
$$\{\pi_A(\sigma_{c=0}(GB_{A, count(distinct\ B)\ as\ c}(R \bowtie_{B=D} S)))\}$$

---

[6] We discuss the issue of dealing with multisets in section 5.5.

(or set intersection instead of join) to get those a values in $R.A$ that are related to values $R.B$ that have no counterpart in $S.D$. However, the expression is clearly incorrect, and will always return the empty relation. The second problem is due to the fact that both the query structure (which set is parametric) and the GQ formula influence the algebraic expression. These two factors may be in collision. To solve these problems, our interpreter proceeds with a case by case analysis. In the following, we pretend (for readability) that all set terms have a single parameter and a single (bound) variable; the approach can be extended easily to the general case. We next present all possible cases and the approaches to evaluate them. For each one of $c_1 = |X_1 - X_2|$, $c_2 = |X_2 - X_1|$, and $c_3 = |X_1 \cap X_2|$, we consider four cases: (a) only the first set term is parametric (b) only the second set term is parametric, (c) both are parametric and share variables, and (d) both are parametric and do not share variables. While each case requires a slightly different approach, the main ideas are the same. For lack of space, we show only our approach on the first case. Thus, assume given a formula $\{x \mid Q(\{y \mid \varphi(x,y)\}, \{z \mid \psi(z)\})\}$, where X_1 is parametric, X_2 is not, and let $cond(c_i)$ be the condition associated with quantifier Q, involving parameter c_i. Assume that the result of E_1 has the attributes A, B, and the result of E_2 has the attribute D, where A is the parameter, and B, D are the bound variables. This $QLGQ$ expression asks for each value of A, the relationship between the sets $\{B\}$ and $\{D\}$.

- **Case 1a** (c_1):
  Computing $c_1 = |X_1 - X_2|$ for case a is equivalent to computing, for each value of A, the number of the attributes which are in $\{B\}$, but not in $\{D\}$. To do so, we perform left outer join of E_1 and E_2, followed by grouping A and computing $count(D)$ when D is null. The formula is:
  $$\pi_A(\sigma_{cond(c_1)}(GB_{A,count(D,\ when\ D\ is\ null)\ as\ c_1}$$
  $$(\sigma_{distinct\ A,B,D}(E_1 \bowtie_{B=D} E_2))))$$
  Note that $\sigma_{distinct\ A,B,D}$ between left outer join and group-by is used to remove duplicates because we will count the number of null values in the next step. Note also that the expression $GB_{A,count(D,\ when\ D\ is\ null)}$ means to count, within each group, how many times we have a non-match. This is not part of standard relational algebra, but it can be easily implementable in SQL with a `CASE` statement. One important point is that left outer join is required if the condition involves $c_1 = 0$; otherwise, we can use antijoin instead and count over B instead of D.
- **Case 2a** (c_2):
  Computing $c_2 = |X_2 - X_1|$ for case a is equivalent to computing, for each value of A, the number of the attributes which are in $\{D\}$, but not in $\{B\}$. It can be obtained by first performing left outer join of E_1 and E_2, followed by grouping A and computing $count(D)$, to compute how many attributes are in common between $\{D\}$ and $\{B\}$; and then computing the number of attributes in $\{D\}$ and using this number to subtract the number in common. The formula is:

$T1 = GB_{A,count(distinct\ D)\ as\ cnt1}(E_1 \bowtie_{B=D} E_2)$

$T2 = AGG_{count(distinct\ D)\ as\ cnt2}(E_2)$

$\pi_A(\sigma_{cond(c_2)}(\pi_{A,cnt2-cnt1\ as\ c_2}(T1 \times T2)))$

Note that in the last expression there is an apparently expensive Cartesian product. Since T_2 has a single row and single column, it is really a trivial operation to implement. One optimization for the above formula is when the condition is $c_2 = 0$, we can use semijoin instead of outerjoin to compute $T1$, and count B instead of D.

• **Case 3a** (c_3):

Computing $c_3 = |X_1 \cap X_2|$ for case a is equivalent to computing, for each value of A, the number of the attributes which are in both $\{B\}$ and $\{D\}$. To do so, we perform left outer join of E_1 and E_2, followed by grouping A and computing $count(D)$. The formula is:

$\pi_A(\sigma_{cond(c_3)}(GB_{A,count(distinct\ D)\ as\ c_3}(E_1 \bowtie_{B=D} E_2)))$

Note that left outer join is needed if the condition involves $c_3 = 0$; otherwise, join can be used instead. Furthermore, by counting over B instead of D, we can convert join into semijoin.

b. X_1 is not parametric, X_2 is:

$\{x \mid Q(\{y \mid \varphi(y)\}, \{z \mid \psi(x,z)\})\}$ Assume that the result of E_1 has the attributes B, and the result of E_2 has the attribute C, D, where C is the parameter, and B, D are the bound variables. This $QLGQ$ expression asks for each value of C, the relationship between the sets $\{B\}$ and $\{D\}$. This case is symmetric to case a.

• **Case 1b** (c_1): symmetric to Case 2a.

• **Case 2b** (c_2): symmetric to Case 1a.

• **Case 3b** (c_3): symmetric to Case 3a.

c. X_1 and X_2 are parametric, share a variable:

$\{x \mid Q(\{y \mid \varphi(x,y)\}, \{z \mid \psi(x,z)\})\}$ Assume that the result of E_1 has the attributes A, B, and the result of E_2 has the attributes A, D, where A is the parameter, and B, D are the bound variables. This $QLGQ$ expression asks for each value of A common between E_1 and E_2, the relationship between the sets $\{B\}$ and $\{D\}$. The difference between case a and case c is that case a computes, for each group of A, the relationship between $\{B\}$ and the whole set of $\{D\}$, while case c computes the relationship between $\{B\}$ and $\{D\}$ based on the same group. Thus, for case c, we need an extra step to obtain the common parameter between E_1 and E_2, then we can use the same formulas and optimizations from case a. • **Case 1c** (c_1):

$T1 = E_1 \ltimes_{E_1.A=E_2.A} E_2$

$\pi_A(\sigma_{cond(c_1)}(GB_{T1.A,count(D,\ when\ D\ is\ null)\ as\ c_1}$

$(\sigma_{distinct\ T1.A,B,D}$

$(T1 \bowtie_{T1.A=E_2.A \wedge B=D} E_2))))$ • **Case 2c** (c_2): symmetric to Case 1c. •

Case 3c (c_3):

$T1 = E_1 \ltimes_{E_1.A=E_2.A} E_2$

$\pi_A(\sigma_{cond(c_3)}(GB_{T1.A,count(distinct\ D)\ as\ c_3}$

$(T1 \bowtie_{T1.A=E_2.A \wedge B=D} E_2)))$ If the condition does not involve $c_3 = 0$, computing T_1 can be avoided and outer join can be replaced by join.

d. X_1 and X_2 are parametric, do not share a variable:
$\{x, w \mid Q(\{y \mid \varphi(x, y)\}, \{z \mid \psi(w, z)\})\}$ Assume that the result of E_1 has the attributes A, B, and the result of E_2 has the attributes C, D, where A and C are the parameters, and B, D are the bound variables. This $QLGQ$ expression asks for each value of A, the relationship between the sets $\{B\}$ and $\{D\}$, where $\{D\}$ is based on each value of C. • **Case 1d** (c_1): Computing $c_1 = |X_1 - X_2|$ for case d is equivalent to computing, for each combination of A and C, the number of the attributes which are in $\{B\}$, but not in $\{D\}$, where $\{B\}$ and $\{D\}$ are based on each value of A and C respectively. To obtain each combination of A and C, we first compute for each distinct value of A, the total number of attributes in E_1, and the distinct values of C in E_2, and then perform a Cartesian product. Note that Cartesian product can not be avoided due to the semantics of this case. What we can do to reduce the cost of Cartesian product is using the minimum size inputs through duplicate removal. Once we have the combination, we first compute how many attributes are in common between $\{B\}$ and $\{D\}$ by performing join of E_1 and E_2, followed by a group-by, and then use the total number to subtract the number in common. The formula is:
$T1 = (GB_{A, count(distinct\ B)\ as\ cnt1}(E_1)) \times \pi_C(E_2)$
$T2 = GB_{A,C, count(distinct\ D)\ as\ cnt2}(E_1 \bowtie_{B=D} E_2)$
$\pi_{A,C}(\sigma_{cond(c_1)}(\pi_{A,C, cnt1-cnt2\ as\ c_1}$
$(T1 \bowtie_{T1.A=T2.A \wedge T1.C=T2.C} T2)))$
Note that, when computing $cnt1 - cnt2$ in the last expression, $cnt2$ may be null. In this case, $cnt2$ should be set to 0, which can be implemented in SQL using a `CASE` statement. If the condition involves $c_1 = 0$, we can perform join instead of outer join on $T1$ and $T2$ on the conditions $T1.A = T2.A$ and $T1.C = T2.C$ and $cnt1 = cnt2$. • **Case 2d** (c_2): symmetric to Case 1d. • **Case 3d** (c_3):
Computing $c_3 = |X_1 \cap X_2|$ for case d is equivalent to computing, for each combination of A and C, how many attributes are common in $\{B\}$ and $\{D\}$, where $\{B\}$ and $\{D\}$ are based on each value of A and C respectively. We distinguish between the case when the condition does not involve $c_3 = 0$ and the case when the condition is $c_3 = 0$. The former case is easy to compute by performing join followed by group-by. The formula is:
$\pi_{A,C}(\sigma_{cond(c_3)}(GB_{A,C, count(distinct\ D)\ as\ c_3}(E_1 \bowtie_{B=D} E_2)))$
The latter case requires a Cartesian product of the distinct values of A and C, whose result is all possible combinations of A and C, and a difference between these combinations and those in common. The formula is:
$(\pi_A(E_1) \times \pi_C(E_2)) - \pi_{A,C}(E_1 \bowtie_{B=D} E_2)$

Algorithm 1 Interpreter I

Require: $QLGQ$ set term S
Ensure: extended relational algebra expression
1: **switch** S **do**
2:    **case** $S_1 \cup S_2$: return $I(S_1) \cup I(S_2)$;
3:    **case** $\{x_1 \ldots, x_n \mid \varphi\}$:
4:       return $\pi_{tr(x_1,\varphi),\ldots,tr(x_n,\varphi) \cup tr(Fvar(\varphi))} IF(\varphi)$;
5: **end switch**

Algorithm 2 Formula Interpreter IF

Require: a $QLGQ$ formula φ
Ensure: extended relational algebra expression
1: **switch** φ **do**
2:    **case** $R(t_1, \ldots, t_n)$:
3:       **switch** t_i **do**
4:          **case** all t_i different variables: return R;
5:          **case** some t_i is a constant c: return $\sigma_{tr(t_1,\varphi)=c}(R)$;
6:          **case** some t_i is the same constant as some t_j:
7:             return $\sigma_{tr(t_i,\varphi) = tr(t_j,\varphi)}(R)$;
8:       **end switch**
9:    **case** $\psi_1 \wedge \psi_2$: return $IF(\psi_1) \cap IF(\psi_2)$;
10:    **case** $\psi \wedge t_1 \theta t_2$: return $\sigma_{tr(t_1,\psi)\theta tr(t_2,\psi)}(IF(\psi))$;
11:    **case** $Q(X_1, X_2)$:
12:       return apply$(Q, I(X_1), I(X_2))$;
13: **end switch**

The final interpreter is shown in Algorithm 1 and Algorithm 2. A database D is fixed and hence not mentioned. We assume a function tr that takes as input a term and a formula and returns, if the term is a constant, the constant itself; and if the term is a variable, the name of the attribute to which domain the variable is grounded in the formula. The function *apply*, in the last line, is assumed to have access to (at least) a PL formula defining each quantifier Q. It simply calls the interpreter recursively on the set terms that are arguments to the quantifier, and applies the transformations described in detail above.

Recall that a query in $QLGQ$ is a non-parametric set term. The following lemma simply states that our interpreter produces correct translations. The proof is omitted due to lack of space.

Lemma 5.3. For query q in $QLGQ$ and database D, $I(g) = \llbracket g \rrbracket_D$. Q.E.D.

Revisiting our old example, the query "find the professors teaching all students", expressed by $QLGQ$ formula
$\{y \mid \textbf{all}(\{x \mid \texttt{student}(x)\}, \{x \mid \texttt{teaches}(y, x)\})\}$
causes the interpreter to attempt to evaluate the formula with the **all** quantifier. This in turn causes *apply* to evaluate the two set terms, which in this case simply return the underlying relations $\texttt{student}$ and $\texttt{teaches}$. Assume

all has been defined as $c_1 = 0$. Since the first set term is not parametric, and the second is, this is Case 1b, which yields the formula:

$T1 = GB_{pid,count(distinct\ sid)\ as\ cnt1}(\text{student} \bowtie_{sid=sid} \text{teaches})$
$T2 = AGG_{count(distinct\ sid)\ as\ cnt2}(\text{student})$
$\pi_{pid}(\sigma_{c_1=0}(\pi_{pid,cnt2-cnt1\ as\ c_1}(T1 \times T2)))$

We emphasize that different definitions of **all** would lead to different formulas, which a cost-based optimizer would consider as alternatives.

Finally, we point out that complex queries (where arguments to a quantifier are in turn queries with quantifiers in them) can be handled by our recursive interpreter without problem; in fact, because any intermediate results are SPJG expressions like the ones seen, any such query can be reduced to a single RA expression and hence optimized as a single block.

5.3.1 Complex Queries

One could have a $LQGQ$ query in which different quantifiers can be used at different levels. As an example, assume the predicate `Friends(st1,st2)` that holds of two students if they are friends of each other. The query *"List the students who have at least 2 friends taught by Peter in some course"* can be expressed in $QLGQ$ as follows:

$\{Sname \mid$ **at least two**
$\quad\quad (\{fr \mid \text{Friend } (Sname, fr)\},$
$\quad\quad \{fr \mid$ **some**
$\quad\quad\quad\quad\quad (\{crs \mid \text{Lectures}(\textbf{Peter}, crs)\},$
$\quad\quad\quad\quad\quad \{course \mid \text{Takes}(fr, course)\})\})\}$

This query could be processed with the approach just described by composition: first, we note that the embedded subquery headed by **some** (the bottom three lines) is just a query of type 3a. Thus, it can be processed normally, and the result is a set term (relation). But if we substitute the bottom three lines by a set term (relation), we again have a normal query of type 3a, which again can be processed normally. The recursive rules that define the syntax of the language ensure that each complex $QLGQ$ query can be broken down in a series of queries that fit into one of the patterns seen, just like this example.

Intuitively, $QLGQ$ formulas can be written as a tree in a manner similar to relational algebra expressions, with basic formulas as leaves, and set terms or quantifiers as inner nodes, and evaluated *bottom-up*, by evaluating all basic formulas in subexpressions, and using the result of such an evaluation to evaluate inner nodes. Note that, thanks to the compositionality of relational algebra, each such complex formula corresponds to a *single* relational expression. This means that an optimizer can decide where (if at all) it is necessary to generate temporary results, and that optimization can be done globally. Thus, in our example, even though we broke down processing into two, a single expression can be generated.

5.4 Optimization

We stress that the quantifiers expressible in PL are all representable in SQL. However, if forced to do so, the SQL programmer must think of how to express the quantifier using aggregates and other features; and the language loses some of its declarativeness. When the desired properties are expressed as quantifiers, it is up to the system to determine what is the best way to implement them. Indeed, one of the most appealing features of generalized quantification is the fact that it is a high-level, declarative approach to query expression. Hence, it leads itself very well to query optimization. In this section, we outline optimization techniques that apply to the expressions produced by the interpreter. For lack of space, we limit ourselves to the most common approaches and to a novel technique based on GQ properties, and do not discuss other approaches that rely on well known relational algebra techniques. We note that the approaches can be combined, as they are orthogonal to each other.

5.4.1 Optimization on RA Expressions

Most formulas introduced in Section 5.3 involve outer joins. Fortunately, as we have pointed out before, outer joins are only required for some special cases where the condition involves comparison with zero. Thus, for other cases, the formulas can be significantly simplified to more efficient expressions. Furthermore, optimization techniques developed for relational algebra, such as pushing down group-by (e.g., [41]) and exploiting *common subexpressions* (e.g., [107, 106]), can be applied to the formulas. Pushing down group-by, whenever possible, can reduce the cost of join or outer join, but it is not always possible. Common subexpressions in $QLGQ$ denotes that the first and the second set terms share common underlying relations and sometimes, common constant conditions. Common subexpressions occur frequently in $QLGQ$; if they are computed once and then reused by later processing, query performance can be greatly improved ([106]).

Another special case which can be further optimized is the case where, for a given $QLGQ$ expression, the bound variable values of the second term are totally included in those of the first term. From the database perspective, this underlying relation denotes have a primary key/foreign key relationship. In this case, join or outer join used to find common attributes is not needed. We only need to count on the bound variables. For instance, in the previous example (end of section 5.3), $T1$ can be simplified as

$GB_{pid, count(distinct\ sid)\ as\ cnt1}(\texttt{teaches})$.

5.4.2 Optimization using GQ Properties

There are several logical properties of quantifiers that can be exploited for optimization. One of them is that of *being a sieve* [16]. The intuition behind this idea is illustrated with an example: consider the GQ $\mathbf{all}(X, Y)$, defined by $X \subseteq Y$. If $X = \emptyset$, then any Y stands in the desired relation $(\mathbf{all}(X, Y)$ is always true) and there is no need to check the condition (hence the value of this property for optimization). We say that $X \neq \emptyset$ is a *sieve condition* for $\mathbf{all}$. By looking at different quantifiers' definitions, it is easy to identify sieve conditions for them. The following definitions and examples formalize the idea and show how it contributes to query optimization. In the following, we consider given a domain M and ignore it in the notation. For quantifier Q, let $Q(X)_1 = \{Y \mid Q(Y, X)\}$ and $Q(X)_2 = \{Y \mid Q(X, Y)\}$.

Definition 5.4.1 Quantifier Q behaves as a sieve in its ith argument $(i = 1, 2)$ whenever for all $X \subseteq M$, $Q(X)_i \neq \emptyset$ and $Q(X)_i \neq \mathcal{P}(M)$. Q behaves as a sieve when it behaves as a sieve in any argument. *Q.E.D.*

Definition 5.4.2 Let $\varphi(X)$ be a formula on set X and Q a quantifier. φ is a *sieve condition* for Q if, whenever $\varphi(X)$ is true, Q behaves as a sieve, that is
$\varphi(X) \to \neg \forall Y \; Q(X, Y)$ and $\varphi(X) \to \exists Y \; Q(X, Y)$ or
$\varphi(X) \to \neg \forall Y \; Q(Y, X)$ and $\varphi(X) \to \exists Y \; Q(Y, X)$ *Q.E.D.*

Note that when a sieve condition fails, two things may happen: all sets qualify for a quantifier's definition, or none does. We distinguish both aspects because they tell us different things about how to proceed in query processing.

Definition 5.4.3 Let $\varphi(X, Y)$ be a formula on sets X, Y and Q a quantifier. φ is an *implication* for Q if, whenever $\varphi(X, Y)$ is true, $Q(X, Y)$ is also true, that is, $\forall X, Y \; \varphi(X, Y) \to Q(X, Y)$. φ is a *constraint* for Q if, whenever $\varphi(X, Y)$ is true, $Q(X, Y)$ is false, that is, $\forall X, Y \; \varphi(X, Y) \to \neg Q(X, Y)$ *Q.E.D.*

GQ Formula	Sieve	Negation
$\|X \cap Y\| > n$	$\|X\| \geq n$ and $\|Y\| \geq n$	constraint
$\|X \cap Y\| < n$	$\|X\| \geq n$ and $\|Y\| \geq n$	implication
$\|X \cap Y\| = n$	$\|X\| \geq n$ and $\|Y\| \geq n$	constraint
$A \subseteq B$	$A \neq \emptyset$; $B \neq M$	implication
$\|X - Y\| > n$	$\|X\| \geq n$	constraint
$\|X - Y\| = n$	$\|X\| \geq n$	implication

Fig. 5.1. Quantifiers, sieve conditions and their negations

Adding a sieve condition to the relational algebra expression for a given quantifier could improve query performance. The sieve conditions for some quantifier formulas are shown in Figure 5.1. As an example, consider the query "select the orders where **all but 10** suppliers are from Europe." The quantifier **all but 10** can be computed by $c_1 = 10$. According to the formulas defined in our interpreter, outer join will be involved. But note that only the orders which have more than 10 suppliers may qualify; others must fail (this GQ corresponds to the bottom line in Figure 5.1). If, before outer join, we add an aggregation to count the number of suppliers for each order and remove orders with less than 10 suppliers, we may reduce the input (and hence the cost) of the outer join. The aggregation is checking the sieve condition. Clearly, adding sieve conditions may not always yield better performance, since checking the condition adds to the cost, and the check may not filter out enough data to justify this added cost. Therefore, a cost-based optimizer should consider both plans (with and without sieve conditions), and choose the most efficient one for a given query.

5.5 Application to SQL

The $QLGQ$ language provides a simple, flexible framework to study quantification without worrying about syntactic details. As an example of the generality of the approach, note that quantification is independent of the data model used; by changing the basic formulas in $QLGQ$ to use XPath expressions, quantification could be applied to semistructured (XML) data (more on this on chapter 9). The relational model has been used as a starting point because it is simple and well known; moreover, one of our goals is to use the lessons learned with $QLGQ$ in SQL. In this section we show how our approach can illustrate some of the weaknesses of SQL with respect to quantification, and how it suggests potential solutions.

As stated above, $QLGQ$ set terms are very closely related to SQL SPJ queries; in particular, set terms used *within* a query correspond to SQL subqueries (parametric set terms correspond to correlated subqueries). We have already shown several examples in section 4.4 of a potential way to incorporate GQs into SQL. One appeal of our approach is that the query structure remains pretty much the same regardless of quantifier used: the query "find the professors teaching half the students" can be expressed as

```
SELECT pid FROM Professor P
WHERE HALF
        (SELECT sid FROM Student)
        (SELECT sid FROM Teaches WHERE pid = P.pid)
```

which is the same query as those of section 4.4 with a different quantifier. Likewise, "find the professors teaching 10%, at least 10, all but 2, ... of the students" would be expressed by simply changing the GQ used. As discussed in section 4.4, an even better approach would be to eliminate the need for correlation by following $QLGQ$ syntax, but this would need further changes to SQL syntax in order to introduce parameters. The following eliminates the FROM clause entirely, by adding a use clause that denotes the parameter:

```
SELECT pid
```

```
WHERE HALF (SELECT sid FROM Student),
           (SELECT sid FROM Teaches USE pid)
```

The advantage of this approach is that it easier to optimize since we do not need to worry about correlations. In fact, this query is so close to $QLGQ$ that it can basically be interpreted in the same manner: the use clause is similar to a GROUP BY and it results on a partition of the table resulting from the select clause by values of pid; then each group in the partition is compared to the relation corresponding to the other set term. Note that this opens up the door to a second implementation, one that joins the relations for each set term and uses a *set predicate* to implement the GQ. This allows us to reuse all the research into implementing set predicates ([72, 84]). The problem with this approach, though, is that it is not as extendible as our approach that uses a PL formula for each GQ and an interpreter that uses the formula to produce an algebraic expression.

Whatever version we use, the point remains that SQL has a mechanism to form sets (subqueries), but that such mechanisms are not fully exploited by the language as only a few predicates (IN, etc.) are allowed to operate on subqueries, and those are *tuple-to-set* predicates.

We can claim that the queries above are easier to write than their standard SQL counterparts. Although *easiness* of use is an informal, difficult to measure property, we believe most SQL programmers would find this query easier than the query in SQL without GQs (expressed either with subqueries that use [NOT] IN, [NOT] EXISTS, or expressed with grouping and aggregation). However, our main point is that by formulating the query with GQs we leave it up to the optimizer to decide how best to implement it: the programmer focuses on the concept; the system has several options to consider the most efficient approach. We stress that [52], which proposes a similar idea, handles only a finite amount of GQs, and does so by translating back into SQL with subqueries (using [NOT] IN, [NOT] EXIST). By contrast, our approach can handle an unbounded number of quantifiers and can do so efficiently.

One of the strong points of our approach is that it allows the addition of particular quantifiers as needed. While it can be argued that users may not be able to express the quantifiers they need as required by our approach, we see this as a task for a DBA-type of user, who can study query workloads and determine if certain frequent patterns can be expressed as a case of generalized quantification. Once an adequate formulation is introduced into the system, the interpreter does the rest. Note that quantifiers, once defined, can be made available to all users and applications accessing the database. Note also that quantifier definitions are portable across databases.

Two issues need to be addressed for quantifiers to become practical operators: handling of duplicates and handling of nulls. So far, we have stressed that we are working with *sets* of elements, since this is what naturally fits with the quantifiers supported, i.e. **at least two** means *at least two different things*. To achieve set semantics, we use projection with duplicate removal and use the aggregate `count` with a `distinct` in its argument. In order to handle multisets, we can adopt multiset semantics for intersection, difference and (more importantly) for cardinality (so that multiset $\{a,\ a,\ b\}$, for instance, has cardinality 3). In practice, this can be achieved by not removing duplicates (in projection or in aggregation). This means that not only we can handle multiset semantics, but it is actually *more efficient* to do so than set semantics. The other practical issue is the nulls. This is a more delicate issue, in that null semantics is an area of still open debate. With respect to our approach, there are two things that must be done. One is to define set operations when nulls are allowed, which can be done by copying SQL semantics as followed in the WHERE clause (since GQs can be seen as complex conditions). In this approach, two nulls are treated as different for computing intersection and difference. The second thing that needs to be done is to use `COUNT(att)` for a given attribute `att`, or `COUNT(*)`, as needed by the semantics, since the former will ignore nulls and the latter count them. Finally, observe that we rely on nulls to mark "lack of match" in outerjoin operations (and antijoin after optimization). If the database contains nulls, it is necessary to use a different marker to denote lack of match in an outerjoin. While less than ideal, this approach can be implemented easily, and is at least as consistent as possible with the SQL approach.

5.6 Monadic vs. Polyadic Quantification

All the previous development has been devoted to monadic quantifiers. In the previous chapter, we used type (1) quantifiers to generate prefixes; here we have limited ourselves to type $(1, 1)$ quantifiers. However, recalling the type hierarchy of Hella described in chapter 3, it is obvious that these quantifiers are very limited in expressive power. In fact, we have seen that we can translate all such quantifiers into SQL. all quantifiers that deal exclusively with sets (the *monadic quantifiers*, of types (1), $(1, 1)$, $(1, 1, 1)$, ...) can be seen along the lines described here as captured by cardinality considerations (and therefore, in finite sets, where all such cardinalities are finite, and hence can be represented by sets of natural numbers). But, while monadic quantifiers are only of limited interest from a theoretical viewpoint (but see [102] for a study of the different expressiveness of some type (1) quantifiers), there are sound reasons to focus the present work on such quantifiers. First, our approach is a practical one, and considerations of efficient implementation are paramount. Beyond monadic, though, complexity raises immediately and strongly (note that quantifiers of type (2) are, essentially, graph properties. Thus, in order to avoid several NP-hard problems, one should carefully choose those type (2) quantifiers that admit feasible implementation. To the best of our knowledge, no such characterization exists yet). Second, even simple set-based computations are not well supported at the language level; [68] shows that there is no efficient way to express most set operations in the relational algebra. Third, type $(1, 1)$ quantifiers offer reasonable practical uses. This is due to a fact familiar to linguists working on formal semantics (see, for instance, [95]). In formal analysis of simple English sentences, GQs play the role of determiners, and set terms play the role of noun and verb phrases. Thus, the pattern $Q(A, B)$ adjusts well to sentence analysis, with Q being a determiner, A a noun phrase (which, combined with Q, gives the subject of the sentence), and B being a verb phrase (which, combined with the subject, gives the whole sentence). Much analysis has exploited this fact and has pointed out that quantifiers of higher types are rare in natural languages. Since querying is (also) a linguistic activity, we believe this gives some (pragmatic) support for the approach proposed here. We will indeed use this line of research extensively in the next chapters.

6

Quantifier Prefixes

6.1 Introduction

In this chapter we restrict ourselves to quantifiers of type (1) and analyze *quantifier prefixes*. We already gave some basic properties for first order prefixes in subsection 2.3.1, but here we will study prefixes with Generalized Quantifiers, as well as other tuples of prefixes not present in *FOL*.

Since we establish in the previous chapter that all monadic quantifiers can be captured by cardinality properties, it would seem that quantifiers of type (1) can be trivially defined. Indeed, the definition 5.1.3 can be applied to these quantifiers and immediately shows that a quantifier $Q_M(A)$ can be defined by two numbers, $|M - A|$ and $|A|$. When the quantifier is *domain independent*, the first of these two numbers is irrelevant, and only $|A|$ matters. Note, though, that not all type (1) quantifiers are domain independent: $\mathbf{all}^{(1)}$ is not; quantifiers like $\mathbf{n\%of}_M(A) = \{A \subseteq M \mid |A| = \frac{|M| \times n}{100}\}$ is not domain independent either. It is arguable that in some of these cases, the relativization to a type $(1,1)$ quantifier, which yields a domain independent quantifier, is more meaningful. But in any case, quantifiers of type (1) can be easily captured by number properties, and hence the approach of the previous chapter applies. Why, then, study such quantifiers separately? The answer is that its type allows such quantifiers to be easily combined into *quantifier prefixes*, and that such prefixes exhibit interesting properties. Prefixes, not isolated quantifiers, are the real theme of this chapter.

Recall from subsection 2.3.1 that all First Order formulas can be written in PNF:

$$Q_1 x_1 \ldots Q_n x_n \varphi(x_1, \ldots, x_n)$$

where each Q_i is one of $\forall$, $\exists$, and φ is quantifier free. The expression $Q_1 x_1 \ldots Q_n x_n$ is the quantifier prefix of the formula. We also pointed out in chapter 3 that GQs of type (1) are the *direct* counterpart to first order quantifiers, in the sense that each GQ takes as argument a set term of the

A.Badia, *Quantifiers in Action: Generalized Quantification in Query. Logical and Natural Languages*, Advances in Database Systems 37, DOI: 10.1007/978-0-387-09564-6_6,
© Springer Science+Business Media, LLC 2009

form $\{x \mid \psi(x)\}$, and hence binds one variable in one formula. Thus, we investigate how queries with prefixes of GQs of type (1) behave in QLGQ.

6.1.1 Linear and Non-linear Prefixes

In subsection 2.3.1 we talked about the dependency between existential and universal quantifiers, captured in Skolem functions. We saw that, basically, every existential quantifier depends on *all* preceding universal quantifiers. That is, in the formula

$$\forall x \forall y \exists z \exists w$$

variables z and w depends on both x and y. This is reflected in the game semantics in the fact that Spoiler would pick a value for x, then a value for y, and then Duplicator would choose values for z and w knowing both choices of Spoiler. But, what if what we wanted to express required z to depend on x *only* and w to depends on y *only*? It is easy to see that there is no way to write a sentence in FO with such a semantics: for all the orderings that we can come up with that have x in front of z and y in front of w will also have both x and y in front of one of z or w. Note that if we have less than 4 quantifiers, we can always order them in a satisfactory way. This is due to the fact that in FO we have 2 types of quantifiers, and they interact one-on-one.

In a seminal paper, Henkin came up with a way to write formulas that express the requirement above. Henkin noted that we write formulas like we write text, from left to right in an ordered way. He then pointed out that there is no reason to do so, and proposed to write formulas like

$$\left.\begin{array}{l} \forall x\ \exists z \\ \forall y\ \exists w \end{array}\right\} \varphi(x, y, z, w)$$

On each line, quantifiers behave normally; in particular, scope is as before -and therefore z depends on x, and w on y. However, there is no scope across the lines, and therefore z depends *only* on x (and w *only* on y). This quantifier prefix goes under the names *Henkin quantifier*, *finite partially ordered quantifiers*, *branching quantifiers* or *nonlinear quantifiers*.

The reason for the name *finite partially ordered quantifiers* can be seen by realizing that in linear prefixes we are imposing a total order $<$ among quantifiers in a prefix, given by their position in the usual left-to-right reading direction. However, in a Henkin quantifier, the order is partial, in that some quantifiers are incomparable -those in different branches. Another way to see Henkin quantifiers is as *dependency relations*[1]:

Definition 6.1. *A Henkin prefix is a triple* $Q = (A_Q, E_Q, D_Q)$, *where* A_Q *and* D_Q *are disjoint finite sets of variables (*A_Q *stands for the universally*

---

[1] The following is from [66], from which most of the material in this section is taken.

quantified variables, and E_Q for the existentially quantified variables), and $D_Q \subseteq A_Q \times E_Q$ is called the dependency relation.

Intuitively, if $(x, y) \in D_Q$, the existential variable y depends on the universal variable x.

We say that a prefix $Q = (A_Q, E_Q, D_Q)$ is a linear prefix if the order relation induced by D_Q: $x < y$ iff $D_Q(x, y)$ is a linear order on $A_Q \cup E_Q$. Another way to visualize the situation is as a graph, where all variables (that is, elements in $A_Q \cup E_Q$) are the vertices, and D_Q is the edge relation. Note that such a graph is a bipartite graph.

We can apply Skolemization also the Henkin prefixes; what is done is to apply the procedure of subsection 2.3.2 to each branch separately. Or, equivalently, we can use another method: define, for $y \in E_Q$, $D(y) = \{x \mid D_Q(x, y)\}$. Then, substitute each existential variable y by a function $f(x_1, \ldots, x_n)$, where $\{x_1, \ldots, x_n\} = D(y)$. When the prefix is linear, this results in the standard procedure. Note, though, that this is a procedure for Henkin *prefixes*, not for arbitrary formulae with Henkin quantifiers. For such formulae, there is no general procedure to apply Skolemization, as there is no prenex normal form for them.

Note that more than two branches and embedded (nested) branches are allowed by the definition. As an example, the following is a valid formulae:

$$\left(\forall x \ \left(\begin{matrix} \exists y \\ \forall z \ \exists w \\ \exists u \end{matrix} \right) \right)$$

Note that this prefix is equivalent to a linear one:
$\exists u \forall x \exists y \forall z \exists w$.

When is a Henkin prefix equivalent to a linear one? The following gives a sufficient condition.

Lemma 6.2. *Given $Q = (A_Q, E_Q, D_Q)$, if for all $y, y' \in E_Q$, $D(y) \subseteq D(y') \vee D(y') \subseteq D(y)$, then Q is a linear prefix.*

What about the Henkin prefixes that are not equivalent to linear ones? What do they exactly add to the language? The answer is, they add a considerable amount of expressive power. As an example, the following sentence (sometimes called the Ehrenfeucht sentence) is true exactly in infinite models:

$$\exists t \ \left(\begin{matrix} \forall \ x \ \exists y \\ \forall \ x' \ \exists y' \end{matrix} \right) ((x = x') \leftrightarrow (y = y') \wedge (y \neq t))$$

To see why, note that Skolemization would substitute y by $f(x)$ and y' by $g(x')$; but since the formula states (in part) that $\forall x \ \forall x'(x = x' \leftrightarrow f(x) = g(x'))$, we can conclude that $f = g$; thus, we can conclude that there is a function f on the domain of the model that is one-one. However, such a function is not onto, as we have $\exists t \ \forall x \ f(x) \neq t$. The existence of such function guarantees that the domain of the model is infinite. But telling the difference between finite and infinite models is beyond the scope of FOL; hence the language obtained with the simple Henkin prefix used in the Ehrenfeucht sentence is already beyond FOL. In general, the following holds:

Theorem 6.3. *Each FOL formula with a Henkin prefix without an even number of negations in front of the Henkin prefix is equivalent to an existential second order formula. Each FOL formula with a Henkin prefix with an even number of negations in front of the Henkin prefix is equivalent to an universal second order formula.*

As a consequence of the famous result by Fagin equating NP and (existential) second order, the logic with (unrestricted) Henkin quantifiers captures NP.

We finish this brief overview answering the following question: Is there a *Normal Form* for Henkin prefixes? It turns out there is: let $\mathbf{H}_k^n$ be the Henkin prefix made up of k different branches (lines), where in each branch there are n universal quantifiers preceding a single existential quantifier. Then, for each Henkin quantifier Q there are natural numbers n, k, such that Q can be rewritten as a $\mathbf{H}_k^n$ quantifier. In fact, in infinite domains Q can be rewritten as a $\mathbf{H}_k^2$ quantifier.

6.1.2 Henkin Prefixes and Generalized Quantifiers

The basic Henkin quantifier H_2^1, that is, a prefix of the form

$$\forall x \ \exists y$$
$$\forall x' \ \exists y'$$

can be captured by the generalized quantifier B of type $[4]$ in domain M, defined by

$$H(R) = \{R \subseteq M^4 \mid \exists f_1 : M \to M \ \exists f_2 : M \to M \ < a, f_1(a), b, f_2(b) >\in R\}$$

The equivalence can be seen by thinking in terms of Skolemization. The generalized quantifier H is usually called the *Henkin quantifier* because of this equivalence.

We already know that this quantifier captures existential second order logic; therefore, it should come as no surprise that many other quantifiers can be expressed with it. As outstanding examples, the Chang, Hartig and Rescher quantifier are all expressible in FOL with H (see chapter 3 for the definition of these quantifiers). What happens when we take $n > 1$ or $k > 2$? As a consequence of the results of Hella ([47]) we know that $FOL(H_k^n)$ is strictly weaker than $FOL(H_{k+1}^n)$ (that is, there are properties expressible in the latter which are not expressible in the former), and also strictly weaker than $FOL(H_k^{n+p})$

for certain p that depends on n and k. As a result, the hierarchy of logics $FOL(H_k^\omega) = \bigcup_n FOL(H_k^n)$ and $FOL(H_\omega^n) = \bigcup_k FOL(H_k^n)$ are *strictly hierarchical*: they contain an strictly increasing (in expressive power), infinite sequence of logics.

Finally, we repeat the observation of subsection 2.3.1, that a FOL sentence can be put into PNF, and the quantifiers in the prefix can be divided into blocks, each block containing all adjacent same quantifiers (all existential or all universal). This partition is basic in the sense that any sentence obtained from a given one by moving quantifiers within a block is equivalent to the original one, while any sentence obtained by moving quantifiers across blocks is different ([100, 63]). These results are extended to the case of generalized quantifiers in [60].

6.2 Linear and Non-Linear Prefixes in $QLGQ$

Let us call $QLGQ^1$ the language obtained by adding to QLGQ only quantifiers of type (1). Under this restriction, the syntax and semantics of quantified formulas are given by the following rules:

- If S is a set term of arity 1, and Q is a GQ of type (1), then $Q(S)$ is a formula in the language.

•

Note that the recursive nature of the definition allows the formation of formulas with quantifier prefixes of arbitrary length. Note also that, by using lifting (see chapter 3), most of what we have to say here (including the definitions above) can be generalized to quantifiers of type (n), for n a natural number, as far as $Q^{(n)}$ is a lifting of a $Q^{(1)}$ quantifier.

It is not difficult to see that, as in first order logic, formulae in $QLGQ^1$ have an equivalent formula in prenex normal form

$$\pm_1(F_1\,\overline{x_1})\,\pm_2\,(F_2\,\overline{x_2})\,\cdots\,\pm_n\,(F_n\,\overline{x_n})\,\phi,$$

where every $\pm_i \in \{\neg, \neg\neg\}$, every $F_i \in \{\exists\}QLGQ^1$, and ϕ is quantifier-free.

As we have already noted in the Introduction, every variable of $\overline{x_i}$ in the formula above is in the scope of the variables of $\overline{x_1}, \ldots, \overline{x_{i-1}}$. The quantifier prefix $\pm_1(F_1\,\overline{x_1})\,\pm_2\,(F_2\,\overline{x_2})\,\cdots\,\pm_n\,(F_n\,\overline{x_n})$ is therefore called *linear*.

We argue, however, that the study of *non-linear* prefixes is interesting for query languages. For example, consider a ternary relation LINEITEM whose tuples are of the form (orderkey, partkey, supkey) relating orders to their ordered parts, together with the parts' supplier. This relation is a simplification of the TPCH benchmark for decision support systems [3]. Now, consider the natural language query "List the orders where (at least) three suppliers supply (at least) five parts". There are at least four possible ways to interpret "three suppliers supply five parts":

1. There are three suppliers that supply five parts each. The parts may be the same or different for each supplier. This reading is expressed in FO(three, five) by

$$\varphi_1(x) := (\text{three } y)(\text{five } z) \text{ LINEITEM}(x, y, z).$$

The quantifier **five** is within the scope of the quantifier **three**[2].

2. There are five parts that are supplied by three suppliers each. The suppliers may or may not be the same for each part. This reading is expressed in FO(three, five) by

$$\varphi_2(x) := (\text{five } z)(\text{three } y) \text{ LINEITEM}(x, y, z).$$

The quantifier **three** is within the scope of the quantifier **five**[3].

3. There are three suppliers and five parts such that each supplier supplies all five parts and all five parts are supplied by each supplier. It is not possible to represent this reading with a linear prefix because here the parts and the suppliers are "picked" from the universe *independently of each other*. By adding branching quantification in which some quantifiers are not within the scope of others, we can express this reading as

$$\varphi_3(x) := \begin{pmatrix} \text{three } y \\ \text{five } z \end{pmatrix} \text{ LINEITEM}(x, y, z).$$

In this formula, each quantifier is independent of each other (being in the topmost or lowermost line has no significance). The interpretation of this formula goes back to [15], which gives the reading indicated below.

4. Finally, it may be that the same three suppliers supply *a total* of five parts among the three of them (i.e., the first supplier supplies two parts, the second supplies two parts more, and the third supplier supplies one more part). This reading can be expressed in FO, but at the cost of quite a bit of complexity:

$$\varphi_4(x) := \begin{array}{l} (\exists y_1) \dots (\exists y_3)(\exists z_1) \dots (\exists z_5) \\ \bigwedge_{(i,j) \in C(3,5)} \text{LINEITEM}(x, y_i, z_j) \end{array}$$

where $C(3, 5)$ denotes all the ways in which the three suppliers can supply the five parts. This formula uses 8 variables, and 21 subformulas. It can be expressed in a shorter way by using *cumulation* quantification:

$$\varphi_5(x) := [\text{three } y, \text{five } z] \text{ LINEITEM}(x, y, z),$$

as we will see below.

---

[2] Note that the statement can be formalized in FO without problems; however, it requires up to 8 variables and a much more complex formula to do so.

[3] Same comments as in the previous case hold.

The important point of our example is that there are readings of some English sentences (expressing queries) which are quite difficult to capture in first order logic because of its combinatorial nature (the cumulative reading), and there are readings which are either difficult or impossible to capture in first order logic (the branching reading). This situation turns out to have a very practical aspect. For the branching readings, those that can be expressed in FO suffer the same inefficiencies in SQL; those that cannot be expressed in FO cannot be expressed in SQL either. For the cumulative readings, SQL is also very inefficient at expressing such readings. Usually, SQL uses its ability to count (to use the COUNT aggregate function, perhaps in combination with its grouping operator) to take care of queries that may be difficult to express in FO. For instance, the query *List the orders with at least 10 suppliers* require a FO formula with 10 different variables and (if implemented literally) 10 joins of the LINEITEM table. However, SQL can simply group by and count to produce a simple, efficient answer. The problem is that such trick is not available for cumulative or branching readings. In the example above, it is not enough to count orders with (at least) 3 suppliers, or orders with (at least) 5 parts. We also need to make sure that those 3 suppliers (among, perhaps, several more) supply those 5 parts (among, perhaps, several more). It is not only necessary to count, but also to keep track of which suppliers and which parts one is counting. This is not possible with current SQL syntax, since attributes that are arguments to an aggregate operator like COUNT cannot be part of the grouping, and vice versa. Therefore, the only way to formulate cumulative queries in SQL is to parallel the FO formulation. Such a query is extremely complex and inefficient (the reader is invited to try to write it down).

Our previous example suggests that branching and cumulation may be of interest in query languages. Let us therefore formally introduce these forms of quantification. We first note that we can view linear quantification as given by formulas of the form

$$\pm_1(F_1 \, \overline{x_1}) \pm_2 (F_2 \, \overline{x_2}) \cdots \pm_n (F_n \, \overline{x_n}) \, \phi,$$

also as an iteration operation *on* generalized quantifiers [101].

Definition 6.4 (Iteration). *If $Q \colon k$ and $P \colon l$ are generalized quantifiers then $Q \cdot P \colon k + l$ is the generalized quantifier*

$$\left\{ R \subseteq \mathcal{U}^{k+l} \mid \left\{ \overline{a} \in \mathcal{U}^k \mid \left\{ \overline{b} \mid (\overline{a}, \overline{b}) \in R \right\} \in P \right\} \in Q \right\}.$$

The reader is requested to verify that $Q \cdot P$ indeed corresponds to the $\mathrm{FO}(Q, P)$ formula $(Q \, \overline{x})(P \, \overline{y}) R(\overline{x}, \overline{y})$. Hence, if $QLGQ^+$ is the closure of G under iteration, then $QLGQ^1 \equiv \mathrm{FO}(QLGQ^+)$, as we can always simulate $(Q \cdot P \, \overline{x}, \overline{y}) \phi$ by $(Q \, \overline{x})(P \, \overline{y}) \phi$.

In a similar vain, we can also define branching and cumulation as operations on generalized quantifiers.

Definition 6.5 (Branching). *If $Q\colon k$ and $P\colon l$ are generalized quantifiers then $\begin{pmatrix} Q \\ P \end{pmatrix}\colon k+l$ is the generalized quantifier*

$$\{R \subseteq \mathcal{U}^{k+l} \mid \text{ there is } S \in Q \text{ and } T \in P \text{ such that } S \times T \subseteq R\}.$$

Example 6.6. The following $\mathrm{FO}\!\left(\begin{pmatrix} \text{three} \\ \text{five} \end{pmatrix}\right)$ formula is equivalent to ϕ_3 above

$$\left(\begin{pmatrix} \text{three} \\ \text{five} \end{pmatrix} y, z\right) \mathrm{LINEITEM}(x, y, z).$$

Also, the three-colorability example from the Introduction can be expressed as

$$\left(\begin{pmatrix} \text{all} \cdot \text{exists} \\ \text{all} \cdot \text{exists} \end{pmatrix} x_1, y_1, x_2, y_2\right) (\exists z_1)(\exists z_2)(\exists z_3)\ \phi_1 \wedge \phi_2 \wedge \phi_3,$$

with all and exists the generalized quantifier forms of $\forall$ and $\exists$; and ϕ_1, ϕ_2, and ϕ_3 as in the Introduction.

In what follows, we will denote the closure of G under iteration and branching by G^b. Hence, $\mathrm{FO}(G^b)$ is the logic where we allow branching of quantifiers in G.

Definition 6.7 (Cumulation). *If $Q\colon k$ and $P\colon l$ are generalized quantifiers then $[Q, P]\colon k+l$ is the generalized quantifier*

$$\{R \subseteq \mathcal{U}^{k+l} \mid \exists S \subseteq R \text{ such that } S \times \pi_{\leq k}(S) \in Q \text{ and } \pi_{>k} \in P\},$$

where

$$\pi_{\leq k}(S) = \{\overline{a} \in \mathcal{U}^k \mid \exists \overline{b} \in \mathcal{U}^l \text{ such that } (\overline{a}, \overline{b}) \in S\} \quad \text{and}$$
$$\pi_{>k}(S) = \{\overline{b} \in \mathcal{U}^l \mid \exists \overline{a} \in \mathcal{U}^k \text{ such that } (\overline{a}, \overline{b}) \in S\}.$$

Example 6.8. The following $\mathrm{FO}([\text{three}, \text{five}])$ formula is equivalent to ϕ_4 above

$$([\text{three}, \text{five}]\ y, z) \mathrm{LINEITEM}(x, y, z).$$

In what follows, we will denote the closure of G under iteration and cumulation by G^c. Hence, $\mathrm{FO}(G^c)$ is the logic where we allow cumulation of quantifiers in G.

We point out that, for simplicity and clarity, the definitions above were for the case of 2 quantifiers. However, all the definitions (iteration, cumulation and branching) can be extended to an arbitrary (but finite) number of quantifiers -and, in the next sections, we will sometimes deal with the case of branching of k quantifiers, for arbitrary k.

At the present, it is not known whether there is an efficient procedure to compute non-linear prefixes, and the answer is very likely negative. Since we already know that branching captures second order logic (and hence NP) we know there is no efficient algorithm for this case. As for cumulation, the issue remains open, but given the combinatorial nature of the definition it also seems unlikely that an efficient procedure can be found. Therefore, in the following sections we restrict ourselves to particular cases of GQs in order to attack the problem of implementation for non-linear prefixes. Next we look at the problem of linear prefixes, and show that a practical implementation actually requires us to deal with some technical issues.

6.3 Cumulation

We now show that cumulation with generalized quantifiers can be expressed in FO extended with the same generalized quantifiers.

Lemma 6.9. *If either Q_1 or Q_2 is empty, then $[Q_1, Q_2]$ is empty.*

Lemma 6.10. *If Q_1 and Q_2 are singletons, then $[Q_1, Q_2]$ can be decided in the relational algebra. That is, there exists a relational algebra expression e that on input R outputs $\emptyset$ if, and only if, $R \notin [Q_1, Q_2]$. By the equivalence of the relational algebra and first order logic, it follows that cumulating singleton quantifiers do not add expressive power to first order logic.*

As a consequence, we have

Lemma 6.11. *If Q_1 and Q_2 are both finite, then $[Q_1, Q_2]$ can be decided in the relational algebra.*

We can build an algorithm to decide to decide $[Q_1, Q_2]$ as follows: since $[Q_1, Q_2] = \bigcup_{k \in Q_1, l \in Q_2} [\{k\}, \{l\}]$, we give a relational algebra expression deciding $[\{k\}, \{l\}]$, First we compute the k-fold Cartesian product of $\pi_1(R)$ and take the Cartesian product of that with R itself:

$$E := \overbrace{\pi_1(R) \times \cdots \times \pi_1(R)}^{k \text{ times}} \times R.$$

Next, select from E those tuples for which the k first columns hold distinct values, one of which is mentioned in the $k + 1$th column. Then project on the final column (which holds the "B" values). Finally, count if the resulting set has cardinality l (this can be done in FO and hence also in the relational algebra).

Finally, $[Q_1, Q_2]$ is decided by $\bigcup_{k \in Q_1, l \in Q_2} e_{k,l}$. Note that this is a valid relational algebra expression since it is a finite union of relational algebra expressions (as Q_1 and Q_2 are finite). let $Q_1 = \{k\}$ and $Q_2 = \{l\}$. The running time in that case is bounded by $O(n^p)$ where $p = \min(\max(Q_1), \max(Q_2))$.

Lemma 6.12 (Range reduction). *Suppose that Q_1 and Q_2 are type (1) quantifiers, with Q_1 finite. Let l be the least element in Q_2 equal to or larger than $\max(Q_1)$ (i.e., $l = \min\{m \geq \max(Q_1) \mid m \in Q_2\}$, assuming such l exists). Let $Q_3 = \{m \leq l \mid m \in Q_2\}$. Then $[Q_1, Q_2] = [Q_1, Q_3]$.*

The above lemma also shows that, when Q_1 is finite and Q_2 is arbitrary (possibly infinite), then membership of $[Q_1, Q_2]$ can be reduced to membership in $[Q_1, Q_3]$ where Q_3 is finite, and for which we can decide membership in the relational algebra.

Furthermore, when both Q_1 and Q_2 are finite, but $\max(Q_1) < \max(Q_2)$, then the previous lemma shows that we only have to check for subsets $S \subseteq R$ up to size $l = \min\{m \geq \max(Q_1) \mid m \in Q_2\}$ (which could be significantly smaller than $\min(\max(Q_1), \max(Q_2))$, the size of the subsets we normally need to check according to the general algorithm).

It is not clear if and how this reduction lemma generalizes to arbitrary cumulations $[Q_1, \ldots, Q_n]$.

Lemma 6.13. *Let Q_1 and Q_2 be type (1) quantifiers that are not disjoint. Let $k = \max(Q_1 \cap Q_2)$ (assuming such k exists), let $Q_1' = \{m \in Q_1 \mid m \leq k\}$, and let $Q_2' = \{m \in Q_2 \mid p \leq k\}$. If $k \neq \max(Q_1 \cup Q_2)$, then $[Q_1, Q_2] = [Q_1', Q_2']$.*

The above lemma shows that, when Q_1 and Q_2 are infinite such that $k = \max(Q_1 \cap Q_2)$ exists, membership of $[Q_1, Q_2]$ can be reduced to membership of cumulation of finite quantifiers. The lemma also shows that, when Q_1 and Q_2 are finite but not disjoint, we only have to check for subsets $S \subseteq R$ up to size k (which could be significantly less than $\min(\max(Q_1), \max(Q_2))$, the size of the subsets that need to be checked according to the general algorithm).

6.4 Branching

The following lemma tells us that we do not need to search for arbitrary elements of $Q_1, \ldots, Q_k$ when we want to verify that a relation $R \in \begin{pmatrix} Q_1 \\ \vdots \\ Q_n \end{pmatrix}$:

Lemma 6.14. *A relation R is in $\begin{pmatrix} Q_1 \\ \vdots \\ Q_n \end{pmatrix}$ if, and only if, for every $1 \leq i \leq k$ there exists $X_i \in Q_i$ with $|X_i| = \min(Q_i)$ such that $X_1 \times \cdots \times X_k \subseteq R$.*

We can use this Lemma to show that branching of type (1) quantifiers does not add expressive power to first-order logic:

Corollary 6.15. *Every branching* $\begin{pmatrix} Q_1 \\ \vdots \\ Q_n \end{pmatrix}$ *with* $Q_1, \ldots, Q_k$ *of type* (1) *is axiomatizable in FO: there exists a FO-sentence* ϕ *over a k-ary relation schema E such that* $R \models \phi \Leftrightarrow R \in \begin{pmatrix} Q_1 \\ \vdots \\ Q_n \end{pmatrix}$.

It readily follows:

Corollary 6.16. *First order logic extended with branching of type* (1) *generalized quantifiers (but not extended with these quantifiers themselves) is equivalent to first order logic.*

Given $Q_1, \ldots, Q_k$ be *fixed* generalized quantifiers. Is there an efficient algorithm to check, given a k-ary relation R whether $R \in \begin{pmatrix} Q_1 \\ \vdots \\ Q_n \end{pmatrix}$? Note that we are interested in data complexity here, not combined complexity as $Q_1, \ldots, Q_k$ are assumed to be fixed.

Let us investigate this question for the setting where $k = 2$ (i.e., R is binary). Let $m = \min(Q_1)$ and let $n = \min(Q_2)$. It follows from Lemma 6.14 that $R \in \begin{pmatrix} Q_1 \\ Q_2 \end{pmatrix}$ if, and only if, there exists a subset $S \subseteq R$ with $|S| = m \times n$; $|pi_1(S)| = m$; and $\pi_2(S) = n$. (Note that for such S we always have $S = \pi_1(S) \times \pi_2(S)$ since always $S \subseteq \pi_1(S) \times \pi_2(S)$ and since the right-hand side has exactly as many elements as the left hand side).

Naive algorithm

Hence, a naive algorithm to decide whether $R \in \begin{pmatrix} Q_1 \\ Q_2 \end{pmatrix}$ would be to consider all subsets S of R of cardinality $m \times n$, and check whether $|\pi_1(S)| = m$ and $|\pi_2(S)| = n$. Let us analyze the running time of this algorithm:

- Given a relation S of cardinality $m \times n$, checking whether $|\pi_1(S)| = m$ and $\pi_2(S) = n$ can certainly be done in time $O((m \times n) \times \log(m \times n))$. Indeed, we simply project on the desired column, eliminate duplicates (using sorting), and count the resulting tuples.
  Since m and n only depend on Q_1 and Q_2, which are fixed, this check is hence done in constant time.
- We need to do the previous step for all $S \subseteq R$ of cardinality $m \times n$. There are at most

$$\binom{p}{m \times n} = \frac{(p - (m \times n) + 1) \times (p - (m \times n) + 2) \times \cdots \times p}{(m \times n)!}$$
$$= O(p^{m \times n})$$

such sets where p denotes the cardinality of R.

We may hence conclude that the naive algorithm runs in time $O(p^{m \times n})$.

A better algorithm

There is, however, a more efficient algorithm to decide $\binom{Q_1}{Q_2}$. In what follows, we assume that $m, n \geq 2$ and that $m \leq n$. We will give an algorithm that decides $\binom{Q_1}{Q_2}$ in time $O(p^m \times \log p^m)$.

Let R be a binary relation. We assume without loss of generality that $|\pi_1(R)| \geq m$ and $|\pi_2(R)| \geq n$. This can clearly be checked in time $O(p \log p)$ and hence also in time $O(p^m)$ for $m \geq 2$. Define the sequence $R_1, R_2, R_3, \ldots$ as follows: $R_1 := R$ and

$$R_{i+1} := \{(a_1, \ldots, a_{i+1}, b) \mid (a_1, \ldots, a_i, b) \in R_i \text{ and } (a_{i+1}, b) \in R\}.$$

Clearly, we can compute R_{i+1} by doing a (modified version of) a nested loop join of R_i and R. Such a nested loop join runs in time $O(|R_i| \times |R|)$. Since R_i contains at most p^i tuples, R_{i+1} can be computed in time $O(p^{i+1})$. In particular, R_m can be computed in time $O(p^m)$.

Now clearly, $R \in \binom{Q_1}{Q_2}$ if, and only if, there exist distinct $a_1, \ldots, a_m$ and distinct $b_1, \ldots, b_m$ such that $(a_1, \ldots, a_m, b_j) \in R_m$, for every $1 \leq j \leq n$. We can check this condition in time $O(p^m \log p^m)$ as follows:

- Fix some arbitrary order $<$ on $\mathcal{U}$.
- Sort R_m lexicographically according to $<$ (i.e.,

$$(a_1, \ldots, a_{m+1}) < (a'_1, \ldots, a'_{m+1})$$

  if there exists some $1 \leq l \leq m+1$ such that $a_{l'} = a'_{l'}$ for every $l' \leq l$ and $a_l < a'_l$). Using standard sorting techniques, this can be done in time $O(p^m \log p^m)$ since the cardinality of R_m is at most p^m.
  Note that, after sorting, all records that agree on the first m components of R_m are grouped together.
- Since R_m is grouped on its first m components, it is easy to check whether there exist distinct $a_1, \ldots, a_m$ and distinct $b_1, \ldots, b_m$ such that $(a_1, \ldots, a_m, b_j) \in R_m$, for every $1 \leq j \leq n$. Indeed, a single scan of the sorted R_m suffices:
  1. Initialize t to the first tuple of R_m
  2. Let $(a_1, \ldots, a_m) := (\pi_1(t), \pi_2(t), \ldots, \pi_m(t))$.
  3. Let $c := 0$
  4. While not exhausted(R) and not $c \geq n$ do:
     - If $(a_1, \ldots, a_m) = (\pi_1(t), \pi_2(t), \ldots, \pi_m(t))$ and $a_1, \ldots, a_m$ are all distinct, then $c := c + 1$ else $c := 1$.

- Let $(a_1, \ldots, a_m) := (\pi_1(t), \pi_2(t), \ldots, \pi_m(t))$.
- let t be the next tuple in R

5. Return $c \geq n$

In conclusion we can compute R_m in time $O(p^m)$, sort R_m in time $O(p^m \times \log p^m)$ and do the final check in time $O(p^m)$. Hence, we can decide $\binom{Q_1}{Q_2}$ in time $O(p^m \times \log p^m)$.

It is interesting to point out that the above algorithm can also be mimicked in SQL. For example, if $m = 3$, then $R \in \binom{Q_1}{Q_2}$ iff the following SQL query does not return an empty answer.

```
SELECT R1.A
FROM
   SELECT R1.A AS A1 , R2.A AS A2, R3.A AS A3, B
   FROM R R1, R R2, R R3
   WHERE R1.A <> R2.A AND R1.A <> R3.A AND R2.A <> R3.A
   AND B IN ( ( SELECT B FROM R WHERE R.A = R1.A)
              INTERSECT
              ( SELECT B FROM R WHERE R.A = R2.A)
              INTERSECT
              ( SELECT B FROM R WHERE R.A = R3.A)
            )
GROUP BY A1,A2,A3
HAVING COUNT(B) >= n
```

6.5 Linear Prefixes

It would seem that linear prefixes are the easy case and should be easy to implement in the appropriate language. However, it turns out that there are some subtleties that make this task quite complicated.

Our goal is to come up with an algorithm expressible in Relational Algebra, perhaps extended with grouping and aggregation, since this is the language underlying SQL and therefore it is known to have efficient implementation. Translation of $QLGQ^1$ formulas would seem straightforward; however, a problem pointed out in the previous chapter will come back to hunt us now. Even though the language we are using is safe and domain independent, it turns out that it does not always deliver the results that one would intuitively expect, as the next example points out.

Example 6.17. We use the relation `LINEITEM(orderkey,partkey,suppkey)` as before. We ask for the orders where no supplier supplies more than five parts. This query can be written as

$\{orderkey \mid$ **no**
$\qquad \{suppkey \mid$ **morethan5**$\}$
$\qquad \{partkey \mid LINEITEM(orderkey, partkey, supkey)\}\}$

However, interpretation of this query yields surprising results. Assume given a database D, and assume that we work under active domain semantics, so any element **a** under consideration comes from the active domain of D. Intuitively, we would expect the above query to "group" the relation LINEITEM by orderkey and suppkey, count how many parts per order and supplier, filter out those that yield 5 or less, and then examine the result per orderkey. If an orderkey was not left with any suppliers after the filter, it would qualify. But in the logical interpretation, for any **a** that does not belong to $\pi_{orderkey}LINEITEM$,

$$D, \mathbf{a} \models \mathbf{no}\{suppkey \mid \mathbf{morethan5}\{partkey \mid LINEITEM(\mathbf{a}, partkey, supkey)\}\}$$

The key issue here is that the quantifier **no** is downward monotone and therefore it admits the empty set as an argument (in fact, its only argument). The empty set represents here the negation of FOL, and Relational Algebra simply does not deal with negation -the most it can do is to deal with set difference, where values are subtracted from a previously constructed set. Intuitively, there is always a domain of discourse that we use as a reference when we issue queries like the one above. The domain of discourse here is the set of all orders; hence, in Relational Algebra a query like the one above would have to be expressed using negation[4]:

$$\pi_{orderkey}(LINEITEM) - $$
$$\pi_{orderkey}(\sigma_{count(partkey) \geq 5}(GB_{orderkey, suppkey}(LINEITEM)))$$

Note that the second set term, to be subtracted from the domain of discourse, is the set of orders such that **there is** at least one supplier that supplies more than 5 parts. The existential quantification is implicit in the semantics of the grouping. This is related to the distinction between the implicit grouping used in correlation and the explicit grouping of the GROUP BY clause, already responsible for the *zero-sum* bug of the original unnesting procedure by Won Kim (REFS).

In order to give intuitive semantics to our example, we can demand that all variables are *grounded*, that is, they take values only from the intended domains in the database. Since each variable in QLGQ is actually bound to a domain in the database, this is perfectly doable. However, it turns out that the

---

[4] Here and in the following we split Relational Algebra formulae in lines for readability, but we emphasize that each formulae is a single expression, which is available to the optimizer/query processor as a single block.

problem is more pervasive than this. Assume that we change the semantics in the language (as we will do next) to make sure that value $\mathbf{a}$ for orderkey only come from $\pi_{orderkey}(LINEITEM)$, and the same is true for other variables (so values for suppkey only come from $\pi_{suppkey}(LINEITEM)$, and values for partkey only come from $\pi_{partkey}(LINEITEM)$). Take $\mathbf{a} \in \pi_{orderkey}(LINEITEM)$ and $\mathbf{b} \in \pi_{suppkey}(LINEITEM)$, but such that they don't appear together in a tuple, i.e. $(\mathbf{a}, \mathbf{b}) \notin \pi_{orderkey,suppkey}(LINEITEM)$. Then the query

List the orders where more than 5 suppliers supply no part.

will yield the $QLGQ$ query

$\{orderkey \mid \mathbf{morethan5}$
$\qquad \{suppkey \mid \mathbf{no}\{partkey \mid LINEITEM(orderkey, partkey, supkey)\}\}\}$

Under the semantics just discussed, any order $\mathbf{a}$ where more than 5 suppliers don't act as suppliers for $\mathbf{a}$ will qualify. However, if we follow the intuitive semantics outlines above, no orders would be retrieved by the Relational Algebra, ever. The problem is that the query above can be understood in two ways: *list the orders where more than 5 suppliers do not act as suppliers for that order*, or: *list the orders where more than 5 suppliers supply no part (whatsoever)*. This second query simply looks for suppliers supplying no part (using set difference) and use this information to select all orders -or none. This is a weird query, but one expressible in relational algebra and SQL.

The moral of the story is that, if downward monotone quantifiers are allowed in the language, linear prefixes are more difficult to evaluate then previously assumed.

Our strategy for evaluating such prefixes is as follows:

1. We ground all variables to make sure that we obtain the intuitive interpretation.
2. We give a relational algebra translation for $QLGQ$ queries that takes into account the intuitive semantics above (we introduce outerjoin for negation in our expressions).
3. We show that the problem above only happens for downward monotonic quantifiers. When no such quantifiers are involved, the $QLGQ$ expression can be simplified considerably, along the lines of the intuitive reading of the formula. Thus, we give a simplified expression for such cases.

6.5.1 Algebraic Translation

Our first step in order to get an effective semantics (i.e. one that can be executed efficiently in a computer) is to restrict ourselves to finite models. Immediately, this implies that we restrict ourselves to finite quantifiers, that is, quantifiers that only have finite sets as arguments. Then we use the argument sketched in the introduction to this chapter, which is an extension of the argument developed in the previous chapter[5]: a GQ Q of type (1) is basically a class of structures of type $< M, A >$, where M is a non-empty set and $A \subseteq M$, that is closed under isomorphisms. In this context, an isomorphism between $< M, A >$ and $< M', B >$ is basically a bijection $f : M \to M'$ such that $f(a) \in B$ iff $a \in A$. Because A and B are sets, this basically comes down to a question of cardinality: if $|M| = |M'|$ and $|A| = |B|$, then $< M, A >$ and $< M', B >$ are isomorphic (note that we restrict ourselves to M, M' finite, so the above implies that $|M - A| = |M' - B|$). Further, when the GQ follows EXT, the only condition that matters is $|A| = |B|$. That is, if $Q_M(A)$ and $|A| = |B|$, then $Q_{M'}(B)$. For reasons explained in the previous chapter, we restrict ourselves to quantifiers that obey EXT. As a consequence, we can identify quantifier Q with an arithmetic formula $f_Q(x)$ with one parameter, such that

$$Q(A) \text{ iff } f_Q(|A|)$$

(recall that we are considering only finite quantifiers!). As an example, $\textbf{all}_M(A)$ iff $|A| = |M|$; $\textbf{some}(A)$ iff $|A| > 0$; $\textbf{at least n}(A)$ iff $|A| \geq n$, and so on.

Next, we point out that $QLGQ$ expressions without quantifiers can be easily translated into relational algebra expressions, as indicated in chapter 4. Our strategy then is as follows: we transform any $QLGQ$ query with quantifier Q into a query in relational algebra where we use grouping and the aggregate function $\texttt{count}$ applied to the relational expression resulting from translating the $QLGQ$ expression without the quantifier. The counting will determine the cardinality of the set associated with a given value (a parameter in the set term), to which a condition can be applied: such condition comes given by f_Q. As an example, the query *List the orders where at least two suppliers supply 4 items each* is written as

$\{orderkey \mid \textbf{atleast2}$
$\qquad \{suppkey \mid \textbf{4}\{partkey \mid LINEITEM(orderkey, partkey, supkey)\}\}\}$

This will become

$$\pi_{orderkey}(\sigma_{ct2 \geq 2}(GB_{orderkey, count(suppkey) \text{ as } ct2}(\sigma_{ct=4}$$
$$(GB_{orderkey, suppkey, count(partkey) \text{ as } ct}(LINEITEM)))))$$

Intuitively, we group by the parameters of the set term, apply $\texttt{count}$ to the bound variable, and obtain a number to which the condition of the quantifier is applied.

However, this approach is only valid when the quantifiers involved are not downward monotone. To deal with these cases, the approach must be extended: following the application of the condition of the quantifier, we must reintroduce the underlying universe of discourse, and pick those values that

---

[5] This approach is already presented in [65].

did *not* qualify. This is done with an outerjoin instead of a set difference. As an example, the query above

$$\{orderkey \mid \textbf{no}$$
$$\{suppkey \mid \textbf{morethan5}\{partkey \mid LINEITEM(orderkey, partkey, supkey)\}\}\}$$

is translated as

$$\pi_{orderkey}(\sigma_{ct2=0}(GB_{orderkey,count(suppkey)\ as\ ct2}$$
$$(LINEITEM \rightthreetimes\!\!\bowtie\!\!\square(\ \sigma_{ct>\ 5}(GB_{orderkey,suppkey,count(partkey)\ as\ ct}(LINEITEM))))))$$

and the query

$$\{orderkey \mid \textbf{morethan5}$$
$$\{suppkey \mid \textbf{no}\{partkey \mid LINEITEM(orderkey, partkey, supkey)\}\}\}$$

is translated as

$$\pi_{orderkey}(\sigma_{ct2\geq5}(GB_{orderkey,count(suppkey)\ as\ ct2}$$
$$(LINEITEM \rightthreetimes\!\!\bowtie\!\!\square(\ \sigma_{ct=0}(GB_{orderkey,suppkey,count(partkey)as\ ct}(LINEITEM))))))$$

Note that we count on the attribute of the right relation; as a result, after the left outerjoin such an attribute will be null for those tuples that do not match (tuples that were eliminated by a previous selection). We are using SQL semantics with respect to nulls, and expect that such a count will return a zero. This allow us to identify the values that were previously rejected.

Finally, we point out that this approach can be used to deal with *domain dependent* quantifiers. These are the quantifiers where we use the underlying domain M in the definition; an example was given before: $\textbf{all}_M^{(1)}(A) = \{A \subseteq M \mid A = M\}$. In our number language, this translates to $\{A \subseteq M \mid |A| = |M|\}$. An example would be a query like *List the orders where all its suppliers supply at least 4 items*, it is understood as comparing the set of all suppliers for a given order (the underlying universe, or M) to the set of all suppliers for the given order that supply at least 4 items (for that order). Here the outerjoin is used to introduce the universe of discourse (M, the set of all suppliers for each order); then, a counter on this outerjoin gives us the term $|M|$ (the total number of suppliers, for each order) used in the definition. On the other hand, the translation of the first quantifier selects, for each order, those suppliers supplying at least 4 items. The selection condition, now, is a comparison of both counters so obtained.

7

Cooperative Query Answering

7.1 Introduction

In this chapter we add a *pragmatic* dimension to our previous analysis. This is consistent with our goal of treating both questions and queries, and their relationship, as a serious and legitimate area of research in databases. Pragmatic issues come about because of the way we *use* natural language to express questions. The claim is that queries (and query languages) that pay attention to such use are more likely to be of relevance to an end user.

This claim is not new; it is the basis for the field of *Cooperative Query Answering* (henceforth CQA). CQA is very diverse; for an overview see [32]. More in-depth information can be found in the book [55] or in the workshop proceedings of [21], [67]. Here we will focus only on the aspects directly relevant to us. Our main point here is that Generalized Quantification is a very useful tool to develop CQA, and we will support this point by showing concrete examples in which GQs illustrate and exemplify CQA techniques.

We will start with a brief overview of CQA for those readers not familiar with the area, and then show how $QLGQ$ fits quite well with the framework.

7.2 Cooperative Query Answering

Cooperative Query Answering may be described as the set of theory, tools and techniques that allow information systems to extend the traditional notion of answer in such a way that the system meets the expectations of intelligent, cooperating agents who try to maximize meaningful information transfer. For instance, in natural language conversations, many implied conventions are followed to ensure a maximum efficiency of communication. These conventions are extremely important, to the point that violating any of them results in the perception, by the other participant in the conversation, that something is wrong. The field of *Pragmatics* is based on the study of these conventions[1].

---

[1] Grice's rules [38] would be a good example of what we have in mind.

A.Badia, *Quantifiers in Action: Generalized Quantification in Query. Logical and Natural Languages*, Advances in Database Systems 37, DOI: 10.1007/978-0-387-09564-6_7,
© Springer Science+Business Media, LLC 2009

The importance of these observations resides in the fact that in order to fulfill the goal of meaningfully communicating with a user, *the database system must be aware of these conventions* so it can act in the cases in which they are violated[2]. We believe that the *incorporation of pragmatics and conversational assumptions into the query language* is a good idea, and we will argue in the next section that this is particularly easy with QLGQ.

Cooperative Query Answering may be composed of one or more of the following [45]:

1. query intent analysis;
2. query rewriting;
3. answer transformation; and
4. answer explanation.

The first stage covers modeling user goals, beliefs, expectations and intentions; we will not address it here. The second stage covers transforming the query based on information about the contents of the database. Most commonly, the techniques in this area are based on interaction with *constraints*, especially integrity constraints (ICs) [33]. Relaxation and generalization are two primary examples of these techniques; in relaxation, the original query is *weakened* so that it admits information that was not originally asked for but is closely related to the information requested. It is usually extremely complex to determine what is the proper way to relax a query, and the analysis mentioned previously about users' goals is used to provide some guidance. Generalization can be seen as a type of relaxation in which weakening is achieved by moving from the concrete to the general in some kind of hierarchy. A framework which uses hierarchies to support query relaxation, generalization and relaxation is presented in [22]; an application of this approach is the CO-BASE system [23], where domain hierarchies are used to guide the manipulation of the query. Another approach to relaxation is that of [31].

The third part covers intensional answers and dealing with presuppositions and assumptions. In intensional answers, a description for the answer is provided instead of (or besides) the *extension* of the answer (i.e. the list of database elements that constitutes the answer set). It is hoped that a short, meaningful description of the answer will be more informative to the user than a list of database elements, especially if the list is long. It also may uncover some users' misconceptions about the database contents. For instance, the query *"Which professors who teach courses to seniors are tenured?"* may be answered with a list of names. However, if *every* professor who teaches courses to seniors is tenured, the answer could be a long list of names, and mere inspection of the list would not reveal that fact (which probably points to a user misconception). Using as an answer a statement of this fact could be more useful to the user than the list of names.

---

[2] This problem also shows up in languages for updating a database.

One of the theses of pragmatics research is that when a user issues a request it is because the user is trying to learn something *new*. It is assumed, then, that the user expects that there are multiple possible answers for her or his query. *Presuppositions* are defined as the set of propositions that the user needs to believe are true in order for the question to have several possible answers (or, equivalently, as the set of propositions such that, if they were false, would imply that the question has only one trivial answer - or none at all). Since we assume that the user wants to maximize the information being conveyed by the linguistic utterances, and that implies asking questions that have several possible answers [38], when a presupposition is violated that fact should be pointed out. Also, when no presupposition is violated but the answer is still trivial, we must look beyond what was explicitly asked and try to find related information that fulfills the user's goal (of acquiring some new knowledge) without being trivial. As an example, if some user asks *"Find the students who have taken all the courses taught by Peter"* it is clear that the user is assuming that Peter has taught some courses.

There is a long research tradition on this topic. Kaplan [58] built a system that used an intermediate layer between the user and the database to deal with presuppositions. He used a representation for queries called MQL (for *Meta Query Language*). MQL represents a query as a graph, where the nodes are sets and the arcs are relations between sets (note the similarity between MQL and the structure of QLGQ queries; the graph structure of MQL fits neatly with the syntax of QLGQ). The system checks that each connected subgraph evaluates to a non-empty answer. If any sub-graph is empty, that is considered a failed presupposition. Janas [57] follows a similar strategy in that he considers all the subqueries of a query. His work is based on the framework of deductive databases; he considers a query as a conjunction of subqueries that must be satisfied. Janas tries to find the smallest subquery of a query that fails, and for that he must consider all subqueries of a query. Janas introduces the concept of *predecessor*, by ordering all subqueries of a query by *logical subsumption* (that is, for formulas φ, ψ, $\varphi \leq \psi$ iff $\models \varphi \rightarrow \psi$). The subqueries of a given query form a lattice with the empty formula at the bottom and the query itself at the top. Since an empty answer to a predecessor implies an empty answer to the query, Janas proposes, in case the query fails, to find the predecessors, sort them, and evaluate them in order. This process will yield the smallest subquery that fails to yield an answer and will, very likely, point out an assumption that does not hold. However, a given query will have many predecessors (for a query with n literals, there will be $2^n - 2$ predecessors, not including the empty predecessor and the query itself). To improve on the situation, Janas proposes several strategies to beat the combinatorial explosion. One of them is similar to the work of Kaplan: two subqueries are defined as *joined* if they share a variable. Two subqueries are *connected* if they are joined, maybe through several subqueries. When searching for failed subqueries, only connected subqueries are considered. Another strategy considers using ICs to eliminate some of the subqueries from consideration. Checking the query against ICs using semantic optimization techniques [19], which are a controlled form of deduction, allows the system to find out if any IC is violated by the query, in which case the query is known to return the empty answer. The work of Gal and Minker [33] is similarly based on checking queries against ICs. This work also uses ICs to detect misconceptions, semantic optimization, and constructing intensional answers. As an example, assume the our database includes the following constraints[3]:

$$\mathbf{all}(\{x \mid \text{Student}(x)\}, \{x \mid \text{Enrolled}(x)\})$$

$$\mathbf{all}(\{x \mid \text{Professor}(x)\}, \{x \mid \mathbf{some}\{y \mid \text{Lectures}(x, y)\}\})$$

Then the query *"Which professors teach classes?"* can be answered intensionally by stating that **all** professors teach some class (the second IC above), while the query *"Is Susan a student?"* can be rewritten as *"Is Susan enrolled?"* given the first IC above.

Finally, *answer explanation* covers creating justifications for answers, that is, explanations of why certain items are (not) part of the answer set, including listing whatever facts and rules were used in constructing the answer set. The ability to justify answers is especially important when the answer obtained is not what the user expected (for instance, an empty answer), or when the answer is so large that its enumeration may overwhelm the user. Seminal work was carried out by Imielinski [54] and extended, among others, by Motro [77]. This is also an area of intense work in the context of deductive databases.

7.3 Cooperative Query Answering with QLGQ

7.3.1 Presuppositions

As pointed out in previous examples, GQs take as arguments set terms that represent NPs in natural language sentences. The existence assumptions of natural language usually imply that the denotation of these NPs (and therefore, of the set terms) is not empty. When this is not the case, the extension of a GQ becomes trivial. Assume, for instance, that the quantifier **all** is given the empty set as its first argument. Then $\mathbf{all}(\emptyset, B)$ is true for any set B (since the semantics of **all** require that $\emptyset \subseteq B$, which holds trivially). It is expected that GQs will divide the domain of discourse into two non-empty regions: that of sets that belong to the GQ's extension, and that of sets that do not. This is called *acting as a sieve* in [16], where it is expected that GQs will act as sieves: "It is often assumed in normal conversation that NPs denote sieves." [16, page 179], and also: "With some NPs it is harder to contradict

---

[3] The following examples are from [32]. The second one requires the use of a *unary* GQ (a GQ with only one argument). **some**(A) is defined as $A \neq \emptyset$.

the assumption that the denotation is a sieve." [16, page 181] (Barwise and Cooper call this kind of quantifier *strong* and the rest *weak*). The importance of this concept lies in the fact that when a GQ does not behave as a sieve, this is a good indication that some presupposition may have been violated.

The preceding intuition can be formalized as follows:

Definition 7.1. *Let Q be a standard quantifier. Then $Q(X)_1 = \{A \mid Q(A, X)\}$ and $Q(X)_2 = \{B \mid Q(X, B)\}$.*

Definition 7.2. *Let Q be a standard quantifier. Q behaves as a sieve in its ith argument ($i = 1, 2$) whenever for all $X \subseteq D$, $Q(X)_i \neq \emptyset$ and $Q(X)_i \neq \mathcal{P}(D)$. Q behaves as a sieve when it behaves as a sieve in any argument.*

Given a quantifier Q, it is possible in many cases to find necessary and sufficient conditions for the quantifier to behave (or not) as a sieve. For instance, **all** does not behave as a sieve if and only if its first argument is empty or its second argument is D; **some** does not behave as a sieve if and only if its first argument is empty or its second argument is empty; and **at least 2** does not behave as a sieve when its first argument or its second argument have cardinality ≤ 1. We call these the *sieve conditions* of the quantifier (many of these conditions are pointed out in [16]).

The idea of sieve condition can be formally expressed, with the above definitions, as follows[4]:

Lemma 7.3. all$(A)_1 = \mathcal{P}(D)$ *iff* $A = D$, *and* **all**$(A)_2 = \mathcal{P}(D)$ *iff* $A = \emptyset$.

The relevance of this concept is that it corresponds to the idea of *degenerate queries* [32], and can be used to formalize the idea of a *violated presupposition*. An example will make this connection clear: to express the query *"find the students who are taking all the courses taught by Peter"*, we use the formula

$$\{Sname \mid \mathbf{all}\ (\{crs \mid \text{Lectures}(\mathbf{Peter}, crs)\},$$
$$\{course \mid \text{Takes}(Sname, course)\})\}$$

If the first set term, which denotes the set of courses taught by Peter, is empty, the query returns all the students that are taking at least one course. To see why, remember that **all**$(A\ B)$ is true iff $(A \subseteq B)$. Thus, if $A = \emptyset$, the formula is true for any B. But note that A being empty means that Peter is not teaching any course, in which case the query makes little sense (most English speakers will agree that, in formulating the query, we are assuming that *there are* courses being taught by Peter)[5].

---

[4] As in standard first order logic, we assume non-empty domains.

[5] Note that this is the same problem that occurs in first order logic with the existential assumption of the universal quantifier, which explains why most logic students, when seeing a formula of the form

$$\forall x\ (P(x) \to Q(x))$$

tend to assume that the antecedent ($\{x \mid P(x)\}$) is not empty.

The sieve concept can be applied to detect mismatches between formal and natural language, especially existence assumptions which are violated. We can make the system aware that GQs are expected to be sieves, and stipulate that it must react when this is not the case. For instance, assume we are processing the above query. As explained before, we first need to determine the extension of the set terms. When $\{crs \mid \text{Lectures}(\textbf{Peter}, crs)\}$ returns an empty result, the application of the GQ **all** will detect a problem, as it will not behave as a sieve (recall that the first set being non-empty is a sieve condition for **all**). From there, the system is able to return to the user a message with *information about what caused the error*. For the user, knowing that the set of courses lectured by Peter is empty (i.e. Peter is not lecturing this semester, for whatever reason), is more informative than a *trivial* or *degenerate* answer which would include every student. Note that if the system had returned such an answer, the user would have no way to *know* that something is wrong (the user could be suspicious of the long list, but it could be concluded that Peter was a really popular professor).

While it is trivial to analyze the **all** case, the advantage of using GQs for this approach is that the same idea can be extended to many other quantifiers. For instance, if the query above asked for all the students who are taking at least two courses from Peter, the query would be expressed in QLGQ with the formula

$$\{Sname \mid \textbf{at least two} (\{crs \mid \text{Lectures}(\textbf{Peter}, crs)\}, \}$$
$$\{course \mid \text{Takes}(Sname, course)\})$$

If Peter had taught only one course that semester, the evaluation of the set term $\{crs \mid \text{Lectures}(\textbf{Peter}, crs)\}$ would return a set of cardinality one, thus violating the sieve conditions of **at least two**. The system then returns a message which points out a *failed presupposition*.

Another, more *active* approach to detect failed presuppositions, is to distinguish the cases in which we do not care about existence assumptions and those in which we do. Thus, for any quantifier Q in the language, we will admit the quantifiers $\boxed{Q}$, $\overline{\underline{Q}}$, and $\overline{Q}$ into the language too. The significance of boxes and half boxes around quantifiers can be readily explained:

- $\boxed{Q}(A, B)$ iff $Q(A, B) \wedge A \neq \emptyset \wedge B \neq \emptyset$;
- $\overline{\underline{Q}}(A, B)$ iff $Q(A, B) \wedge A \neq \emptyset$; and
- $\overline{Q}(A, B)$ iff $Q(A, B) \wedge B \neq \emptyset$.

For instance, we know by [16] that **all** is a strong quantifier, and so we extend it to a $\boxed{\textbf{all}}$ by taking care of the case in which it behaves improperly (i.e. the case in which its first argument is empty). Thus we have *two* different quantifiers, **all** and $\boxed{\textbf{all}}$; by taking a look at their associated interpretation we see that the difference is precisely that $\boxed{\textbf{all}}$ is undefined when its first argument is empty. Considered as functions, we can say that **all** is total and $\boxed{\textbf{all}}$ is partial; this allows us to catch exceptions by extending the partial function to return an error value in the relevant cases. So the proper way to formulate the query above is as follows:

$$\{Sname \mid \boxed{\text{all}}\ (\{crs \mid \text{Lectures}(\textbf{Peter}, crs)\},\ \{course \mid \text{Takes}(Sname, course)\})\}$$

Similar extensions may be applied to other quantifiers such as **no**, **some**, **at least two** etc. to detect explicitly the case of empty argument(s). This pragmatic approach allows us to include a great amount of information with a minimum of symbols[6].

In closing this subsection, we point out another technique that could be improved in the context of QLGQ. Checking the subqueries of a query to detect violated presuppositions is a technique developed in previous research ([58], [57]). However, these approaches have to evaluate a high number of subqueries, as the subqueries of a query form, when ordered by inclusion, a full lattice with an exponential number of elements. With the QLGQ query, instead of a full lattice we have a reduced number of subqueries. We argue that the subqueries generated by a QLGQ query are those that reflect the user assumptions and therefore this method helps focus on the relevant subparts of the query. For instance, in the examples above there are only two subqueries of interest: the set of courses taught by Peter, and the set of courses taken by each student. As queries become more complicated, QLGQ syntax ensures us that the number of subqueries grows linearly with respect to the number of GQs used in the query. For instance, in example 7.3.1 we have a $QLGQ$ query which uses two GQs and has three subqueries. By contrast, the same query can be expressed in FOL as

$$\{fr \mid \exists x\ \exists y\ \exists crs\ \exists crs_2\ Friend(fr, x)\ \wedge\ Friend(fr, y)\ \wedge\ Takes(x, crs)$$
$$\wedge\ Takes(y, crs_2)\ \wedge\ Teaches(\textbf{Peter}, crs)\ \wedge\ Teaches(\textbf{Peter}, crs2)\}$$

This formula contains six atomic subqueries, which can be combined in a lattice of 2^6 elements. Eliminating non-connected subqueries would help, but it still does not reduce the number of subqueries to be considered to the linear number of the QLGQ formula.

7.3.2 Constructing Explanations and Justifications

Another effect of the structural simplicity of QLGQ queries (and their evaluation procedure) is that we can obtain *explanations* for the answers generated against a database in a very simple manner. That is, we can answer the question: Why does this or that value (not) appear in the answer set? This is

---

[6] In contrast, a first order logic formulation would have to make the semantic constraints explicit:

$$\{sname \mid \exists\ c\ \text{Lectures}(\textbf{Peter}, c) \wedge \forall\ x\ (\text{Lectures}(\textbf{Peter}, x) \rightarrow (\text{Takes}(sname, x)))\}$$

achieved by tracing through the evaluation of the QLGQ query with the given value, and checking for the sieve conditions of the GQs used. Using again example 7.3.1 (which asked for the students who have at least two friends taught by Peter in some course), a student may not qualify for the answer to this query for a variety of reasons: he/she has no friends, or he/she has friends but they take no classes, or the classes those friends take are not taught by Peter. Given any student value, we can determine the exact reason he/she is not in the answer set. Also, for elements in the answer set we can determine *why* they are in the answer set, i.e. who are the friends that are taught by Peter.

Assume that Joe is a student. Following the query tree in a top-down manner, we can check the following conditions:

1. Does Joe have at least two friends (i.e. does $\{fr \mid \text{Friends}(\mathbf{Joe}, fr)\}$ respect the sieve conditions for **at least two**?)
2. If Joe has at least two friends, say Carol and Ted, do those friend take any courses (i.e. do $\{course \mid \text{Takes}(\mathbf{Carol}, course)\}$ and $\{course \mid \text{Takes}(\mathbf{Ted}, course)\}$ respect the sieve conditions for **some**?)
3. Does Peter teach at least one course (i.e. does $\{crs \mid \text{Lectures}(\mathbf{Peter}, crs)\}$ respect the sieve conditions for **some**?)
4. If the friends do indeed take courses, are they courses that Peter Lectures (i.e. what is the relationship between $\{course \mid \text{Takes}(\mathbf{Carol}, course)\}$ and both $\{course \mid \text{Takes}(\mathbf{Ted}, course)\}$ and $\{crs \mid \text{Lectures}(\mathbf{Peter}, crs)\}$)?

If Joe is a part of the answer, by following the procedure we can determine:

1. which friends helped Joe qualify for the answer;
2. which courses each one of these friends take; and
3. which of those courses are taught by Peter.

On the other hand, if Joe is not part of the answer, then it is the case that one of the conditions in the list above is not met. This condition will constitute an explanation of Joe not being part of the answer. The subtle improvement that GQs bring in is the fact that we can distinguish between different types of explanation, in the following sense: it is possible that no sieve condition is violated (Joe has at least two friends, those two friends are taking some course, Peter is lecturing at least one course), and yet Joe fails to qualify for the answer, because the relations demanded by the quantifiers do not hold. An explanation is also produced for this case, but this failure is dependent on the state of the database at the time of querying, and therefore is conceptually different from the failure due to the lack of sieve conditions, which denote some presupposition violation. Thus, explanations can point out when a failure is normal and when it is abnormal.

The previous examples can be computed as follows (we sketch an algorithm with pseudo-code and an example to give an intuition of the process). We use Justify to explain why a certain data element $\bar{a}$ is in the answer set of a query represented by tree n, and Explain to point out the reason that the query returned an empty answer. Both algorithms assume that the query is answered following the bottom-up approach algorithm of chapter 8, and that the partial answers for every node in the tree are kept (we call $Answer(n)$ the answer for node n).

Algorithm 3 Justify(n:node, $\overline{a}$:data)

---

if n is a node of type $\wedge, \vee, \neg$ **then**
    call Justify(left(n), $\overline{a}$), Justify(right(n), $\overline{a}$)
end if
if n is a GQ **then**
    mark the sets in $Answer(right(n))$ and $Answer(left(n))$ that pass the quantifier test;
    call Justify(left(n), $\overline{a}$), Justify(right(n), $\overline{a}$)
end if
if n is of type set **then**
    if n has a non-empty free variable list **then**
        let F be the free variable list of n;
        for all $v \in F$ **do**
            call Justify(left(n), b), where b is the value of $\overline{a}$ given to v in $Answer(n)$
        end for
    else
        /* n has an empty free variable list */
        let B be the bound variable list of n;
        for all $v \in B$ **do**
            call Justify(left(n), b), where b is the value given to v in $Answer(n)$
        end for
    end if
end if
if n is a leaf **then**
    in $Answer(n)$ bind the values passed in $\overline{a}$ to their respective variables;
    return the values for other variables.
end if

Because of the constraints of the language, if n is a set node with non-empty free variable list, then $Answer(n)$ contains a relation of the form $\{< \overline{a}, S_{\overline{a}} >\}$. Some of the sets $S_{\overline{a}}$ have been marked in a previous step as contributing to the solution. We call recursively with the values in those sets as parameters.

In the query of example 7.3.1, a call to Justify with the value **Joe** would result in the node **at least two** marking on both subnodes the sets that did qualify for the answer and calling recursively with the value **Joe** to both left and right subnode. The right subnode would pick, among those sets S_{Joe} associated with the value **Joe**, the ones that were marked in the previous step, and for each value b on it call the subtree, which would simply return the fact Friend(b, **Joe**) (indicating that b is a friend of **Joe** who helped qualify him for the answer). On the right subnode the set node would call recursively with all values of fr in its answer set instead of **Joe**, and the leaves would return the set of students who take courses from **Peter**.

The algorithm Explain is based on the fact that in computing an answer to a QLGQ query, we generate values from the interpretation in the leaves, and then filter those values as we go up the tree. Therefore, if a query returns an empty answer it is because at some point in the process a tree node returns an empty result but receives non-empty results as input (or some leave(s) do not generate values to start with). From that point up, results are empty; therefore we look for one such node. Furthermore, in doing so, we can ignore set nodes, since if a set node **n** returns an empty result, this must be because $left(\mathbf{n})$ provides an empty result, as set nodes only pass results along or group results received. Thus, there are only two situations that we must take care of:

1. a leaf that generates an empty result. This indicates a *user misconception* about the database structure or contents.
2. a quantifier node that received non empty results but generates empty results. This in turn may be due to two reasons: the sieve conditions are violated or the sets simply do not meet the requirements of the quantifier.

In the query of example 7.3.1, a call to Explain would result in going down the tree until a node with non empty results as input is found. If such node is the node **at least two**, we would check its sieve condition (that both input results have cardinality ≥ 2). If it was found out that the sieve condition was violated, an appropriate message would be generated in line 21. But if the sieve conditions were not violated, then it is simply the case that the data in the database does currently give an empty answer to this query (hence the message in line 23). If the loop does not stop until a leaf, for instance, Friend($Sname, fr$) (which implies that this atomic formula returns an empty answer), we must conclude that such database relation is empty or does not exist, which arguably pinpoints to a user misconception (hence the message in line 12).

7.3.3 Relaxed Queries

The idea behind *relaxed queries* is to expand the original query in order to retrieve related, relevant information, in order to give the most informative answer. In QLGQ there are two ways to relax a query: the first one is to relax the GQ used, the second one is to relax the set terms. We give a brief example of each technique.

For relaxing the GQ, assume the query *"Are all students enrolled?"*, expressed as follows in QLGQ:

$$\{() \mid \mathbf{all}(\{x \mid \text{Student}(x)\}, \{x \mid \text{Enrolled}(x)\})\}$$

Algorithm 4 Explain(n:node)

---

if **n** receives empty relations as input and **n** is not a leaf **then**
    call Explain(left(**n**));
    if there is a right(**n**) **then**
        call also Explain(right(**n**));
    end if
end if
if **n** is a set node **then**
    call Explain(left(**n**));
end if
if **n** is a leaf node **then**
    if **n** generates an empty answer **then**
        return a *misconception about the database* message and stop.
    end if
end if
if **n** is a node of type $\wedge, \vee, \neg$ **then**
    call Explain(left(**n**));
    call Explain(right(**n**)).
end if
if **n** is a GQ Q **then**
    if no set passes the sieve conditions for Q **then**
        return a *presupposition violated* message and stop.
    else
        return a *not right data in the database* message and stop.
    end if
end if

The quantifier **all** can be computed as follows: **all**(A, B) is true if $|A \cap B| = |A|$. Suppose that in the course of computing the answer, we find out that actually $|A \cap B| = |A|/2$. Then the answer is *no*, but we also know from our computations that half of them are (since $|A \cap B| = |A|/2$ corresponds to the quantifier *half of As are Bs*). Correspondingly, if the final result of the computation is $|A \cap B| = |A|/3$, then we know that the answer is *No, but a third of them are*, and if $|A \cap B| > 0$, then the answer is *No, but some (a few) are*.

The second technique, relaxing the set terms, has been exploited before in the research literature. The novelty here is that we can determine when it is a good idea to do so, based solely on the structure of the query. Since the answer depends not only on the set terms involved but on the quantifiers used, we have to determine what the effect of relaxing the set term is going to be on the query. This can be achieved by using a property of GQs that we define next:

Definition 7.4. *A standard GQ Q is* upward monotonic *in the first argument if whenever $Q(A, B)$ and $A' \subseteq A$, it is the case that $Q(A', B)$. A standard GQ Q is* downward monotonic *in the first argument if whenever $Q(A, B)$ and*

$A \subseteq A'$, it is the case that $Q(A', B)$. The same definition applies to the second argument.

A quantifier may be upward monotonic in one argument and downward monotonic on the other. We denote upward monotonicity by using a $\uparrow$ and downward monotonicity by using a $\downarrow$ and putting them in front (behind) of the symbol MON to denote the behavior of the first (second, respectively) argument. For instance, **all** is $\downarrow MON \uparrow$, **some** is $\uparrow MON \uparrow$, and **no** is $\downarrow MON \downarrow$.

A simple example will show the utility of the concept: assume the query is *"List the professors who teach at most two students with a GPA of 3.0 who are friends of Peter"*. This query would be written in QLGQ as

$$\{pr \mid \textbf{at most 2} \ (\{st \mid \text{Teaches}(pr, st) \land GPA(st, 3.0)\},$$
$$\{st \mid Friend(st, \textbf{Peter})\})\}$$

Suppose that the answer to this question in the current state of the database is empty, and we decide to relax the question. For that, we may try to change the set term $\{st \mid \text{Teaches}(pr, st) \land GPA(st, 3.0)\}$ to $\{st \mid \text{Teaches}(pr, st)\}$, in effect dropping the requirement that students have a certain GPA. The problem is that since **at most two** is a $\downarrow MON \downarrow$ quantifier, relaxing the sets will not help get any answers. To see why, recall that **at most two**(A, B) is defined as $|A \cap B| \leq 2$. Indeed, if A and B are such that this does not hold, any superset of A will not qualify either, even if B is kept fixed, and the same applies to supersets of B. But relaxing a set means ending up with a superset of the original. On the other hand, a subset of A (or B) may qualify. Hence, it may be a good idea to *tighten* the set term by adding predicates that may yield a subset of the original extension. Thus, the heuristic rule is: *when trying to relax a query with GQs that returned an empty answer, relax the set terms that are arguments in an upward monotone position to the quantifier, and tighten the set terms that are arguments in an downward monotone position to the quantifier*. Thus, if the query were *"List the professors who teach all students with a GPA of 3.0 who are friends of Peter"*, we could write the following QLGQ query:

$$\{pr \mid \textbf{all} \ (\{st \mid \text{Teaches}(pr, st) \land GPA(st, 3.0)\},$$
$$\{st \mid Friend(st, \textbf{Peter})\})\}$$

If this query returns an empty answer and we want to relax it, and since **all** is $\downarrow MON \uparrow$, we have two choices: to tighten the set term that is the first argument to the quantifier, or to relax the set term that is the second argument to the quantifier. Thus, both tightening the first set (which amounts to restricting the set of students under consideration) or relaxing the second (which amounts to considering more people besides Peter's friends) should help get an answer. On the other hand, if the original query returned an empty answer (i.e. nobody teaches all students with a GPA of 3.0 who are friends of Peter), it is clear that relaxing the first set (i.e. considering even more students) will not help find answers.

Note that this technique can be seen as an alternative to relaxing the quantifier, as done previously. Thus in this example, if we are only interested in Peter's friends who are students with a 3.0 GPA, and nothing more (or less) will do, we may settle for people who teach two thirds of such students or more (i.e using the GQ $\geq 2/3$ instead of **all**[7]). In any case, the user may choose the option which better suits his or her purposes.

Before converting the heuristic given above into an algorithm, a subtle point regarding quantifier interaction must be discussed. Assume the query *"List the students who have at most 2 friends taught by Peter in at most one course"*, which is written in QLGQ as follows:

$$
\{ Sname \mid \textbf{at most two}
$$
$$
(\{ fr \mid \text{Friend} \qquad (Sname, fr) \},
$$
$$
\{ fr \mid \textbf{at most one}
$$
$$
(\{ crs \mid \text{Lectures}(\textbf{Peter}, crs) \},
$$
$$
\{ course \mid \text{Takes}(fr, course) \}) \}) \}
$$

If this query returns no answer, and we try to relax it, we should actually tighten the two sets that are arguments to **at most two**, as discussed previously. However, one of them is composed of another quantifier, **at most one**, and two set terms that are arguments to **at most one**. Since **at most one** is of type $\downarrow MON \downarrow$, to tighten that set term actually means to *relax* the arguments to **at most one**, since if we tighten the arguments of a downward monotone quantifier, we risk generating *more* results, i. e. relaxing the set term to which the GQ belongs. In the current example, if we relax $\{ crs \mid \text{Lectures}(\textbf{Peter}, crs) \}$ or $\{ course \mid \text{Takes}(fr, course) \}$, we could end up generating more value for fr because of the downward monotonicity of **at most one**. Thus, the algorithm must deal with possible changes of strategy during query analysis. Therefore, we use a parameter, *strategy*, which can be set to one of two values: *relax* and *tighten*. We also define the functions *set-strategy-left* and *set-strategy-right* that take care of setting the parameter to avoid the situation described above.

We call the following algorithm Rewrite. Again, we show an sketch of the procedure in pseudo-code, not a completely formal algorithm.

The auxiliary functions are easy to express, and are shown next.

7.3.4 Expressing and Using Constraints

As explained in subsection 3, the expressiveness of QLGQ depends on the set of GQs that is used with the language. The GQ concept is a very powerful one; in particular, GQs are able to express complex constraints which are hard

---

[7] $\geq 2/3(A, B)$ is easily definable as $|A \cap B| \geq \frac{2}{3} \times |A|$.

Algorithm 5 Rewrite(n:node, strategy:parameter)

---

if n is a set node **then**
    call Rewrite(left(**n**), strategy);
end if
if n is a leaf node **then**
    use standard techniques to apply *strategy* to **n**
end if
if n is a node of type $\wedge, \vee, \neg$ **then**
    call rewrite(left(**n**), strategy);
    call Rewrite(right(**n**), strategy). /* Except for $\neg$ */
end if
if n is a GQ Q **then**
    call rewrite(left(**n**), set-strategy-left(Q, strategy));
    call Rewrite(right(**n**), set-strategy-right(Q, strategy)).
end if

Algorithm 6 Set-strategy-left(Q:GQ, st:parameter) returns parameter

---

if Q is $\uparrow MON$ and st $=$ *relax* **then**
    return *tighten*;
end if
if Q is $\downarrow MON$ and st $=$ *tighten* **then**
    return *relax*;
else
    return st;
end if

Algorithm 7 Set-strategy-right(Q:GQ, st:parameter) returns parameter

---

if Q is $MON \uparrow$ and st $=$ *relax* **then**
    return *tighten*;
end if
if Q is $MON \downarrow$ and st $=$ *tighten* **then**
    return *relax*;
else
    return st;
end if

---

or impossible to express in first order logic. In particular, it was stated that QLGQ with the quantifiers **some** and **no** is equivalent in expressive power to first order logic[8]. While addition of first order definable quantifiers (like **all**, **not all**, **at least two**) does not change the situation, it is relatively easy to define GQs (for instance, **H**), that take the language beyond that level of expressiveness. For example, the constraint *"There are exactly five professors who teach all students"*, can be formulated as:

---

[8] A formal proof can be found in [7].

five $(\{p \mid \quad \text{Professor}(p)\},$
$\qquad \{q \mid \textbf{all} \ (\{x \mid \text{Student}(x)\},$
$\qquad \qquad \qquad \{y \mid \text{Teaches}(q, y)\}))$

This very same constraint is expressible in first order, but is quite tedious and awkward to write down. On the other hand, the constraint *"all teachers teach the same number of students"*, expressible in QLGQ as:

$$\textbf{all} \ (\{x, z \mid \quad Professor(x) \wedge Professor(z)\},$$
$$\{x, z \mid \ \textbf{H} \ (\{y \mid Teaches(x, y)\},$$
$$\{y \mid Teaches(z, y)\}))$$

is not first order definable. Therefore QLGQ can, in combination with the right set of GQs, express a rich set of constraints. Many of the methods used in Cooperative Query Answering rely on having knowledge about the semantics of the data, which is usually expressed (especially in the relational model) through constraints and rules. For instance, integrity constraints may be used to rewrite queries (for simplification or relaxation), or to build intensional answers. Therefore, a richer set of constraints leads to a potential for further usage of these methods.

A richer set of constraints also opens up the possibility of richer deductive processing. Assume the following rule:

$$\textbf{25\%} \ (\{x \mid Student(x) \wedge Status(grad, x) \wedge Major(x, CS)\},$$
$$\{x \mid Student(x) \wedge GPA(x, y) \wedge y > 3.5\})$$

which expresses the fact that 25% of the grad students in CS have a GPA of 3.5 or greater.

If we are given the query *"Find how many graduate students there are in Computer Science with a GPA of 3.5 or better"*, then we can answer the question indirectly (by stating that 25% of them do) or we could rewrite the query to find *all* students in Computer Science and then multiply that number by 0.25. But if the question were *"Find how many graduate students there are in Computer Science with a GPA less than 3.5"*, we could deduce that the answer is 75% of them, and again respond indirectly or directly using the same approach as before. The ability to produce this kind of reasoning implies a deduction system for finite cardinalities, which to our knowledge has not been fully developed yet.

7.4 Further Research in CQA

The use of Generalized Quantifiers in query languages is an effective tool in designing query languages that are responsive, allow for easy expression of complicated queries, and support the kind of rich reasoning needed for Cooperative Query Answering in a natural way. We show how Generalized Quantification can be used in dealing with false presuppositions, constructing justifications and explanation, and query relaxation. Clearly, further research is still necessary. In particular, we note that it is possible to define the language $QLGQ$ on top of other data models with just some minor changes, since the concept of GQ is independent of the data model (this option is explored in section 9.3). Therefore, the techniques explored here could be applied in a larger context than that of the original research. It is interesting to note that, with more complex data models, and models that allow more degrees of freedom, the possibility for misunderstandings grow, and therefore the need for CQA methods is even more acute. To focus on the context of semistructured data (XML, HTML), we note that in such a model the user is expected to specify *paths* to navigate through complex data elements. This is usually done using a language like XPath, which allows path expressions that include wildcards, conditions, and other constructs. However, not all elements are required to have the same schema. Therefore, in semistructured data there are several reasons for failure to obtain any data in an answer: a given path may fail to exist at all in a database (denoting a user misunderstanding about the data and its structure), or the path itself may exist, while no data in the database fulfills all other conditions that are attached to the path. Even if path expressions denote some database elements, the conditions posed by the quantifier may not be met. It is therefore very important to establish exactly the cause of a failure in the context of semistructured data. The exact strategy to *repair* the failure depends on this cause. Allowing partial matches against a path expression may address failures caused by data that deviates from the expected structure, but strategies like relaxation may be needed if the failure is due to available data sets failing to meet the condition imposed by a quantifier. Finally, completely new techniques may be developed in this context. For instance, in an XML database it may be possible to generate sets of complex elements, such that the elements of the set have irregular structure. Modifying the elements as needed to make comparison with other elements meaningful would help maximize the use of set-oriented operations.

8

Generalized Quantifiers and Natural Language

8.1 Introduction

In this chapter we continue the *linguistic turn* that we initiate in the previous one by addressing pragmatic issues. Here we deal squarely with linguistic issues, that is, with questions instead of queries. We can do this because one of the appeals of the concept of GQ is that it has been widely used in linguistic analysis. In particular, the notion of interrogative quantifier has been proposed and analyzed ([94, 20, 42, 35]. Here, we build on such analysis to show how QLGQ can be used to formalize questions, thus providing a solid formal framework for work in *Question Answering* (henceforth QA).

We first provide a (very brief) overview of QA, in order to make the material as self-contained as possible. We then show in section 8.3 how GQs have been used by linguists for the formal analysis of natural language. We then show how $QLGQ$ can be used to support QA. The approach presented here, however, opens up more questions than it answers, and hence we close the chapter with a detailed discussion of our motivation, and where to go from here.

Section 8.3, while too brief, will serve to illustrate how the concept of GQ has been studied and used in several fields, most of the time in isolation from related work in other fields. And yet, we see the same basic concepts in action. Hopefully this brief exposure will entice some researchers to look beyond their own field and facilitate inter-disciplinary research.

8.2 Question Answering

While searches in IR systems usually take in a few keywords and return a ranked list of documents, sometimes the questions are complete sentences and the answer is a piece of information. For instance,

- What is the height of Mount Everest?

A.Badia, *Quantifiers in Action: Generalized Quantification in Query. Logical and Natural Languages*, Advances in Database Systems 37, DOI: 10.1007/978-0-387-09564-6_8, © Springer Science+Business Media, LLC 2009

- How many liters are there to a gallon?
- Who is the most frequent coauthor with Gerald Salton?

A *Question Answering (QA)* system is a system designed to answer such questions. Clearly, to answer questions a QA system needs not only a collection of documents, but possibly auxiliary knowledge bases, like lexicons, ontologies, etc. and may also need to be able to do some inferencing. Some questions may be easy to answer, in the sense that what is asked is the value of a property of an entity (as in the first example above), so that this is similar to slot-filling in IE systems. However, there are more open-ended questions whose answer may not directly be found on a piece of text. Thus, building a QA system, especially an open domain one (opposed to a restricted domain one) may be very difficult.

QA systems usually build an inverted index of the document collection, just like IR systems. However, unlike IR systems, the questions need to be parsed. Once parsed, the question may be classified as one of several types. Associated with each type there is a type of answer (a number, for the first and second questions above; a person name, for the third). Then words appearing in the question and words suggested by the answer type are used to compose a keyword search, handled as in IR. Once a ranked list of documents is returned, the QA system examines each document and breaks it down into segments -QA systems assume that answers will be contained in a small segment of a document. Segments are chosen so that they contain all/most of the query terms and are delimited by sentence or paragraph boundaries. Then each segment gets a score with respect to the original query. Usually this score is based on the number of query terms appearing, or their density (i.e. the proportion of query terms to all terms in the segment). Note that the scoring may be done with some reasoning and knowledge; if the question is "What is the color of grass?" and a sentence in a document is "The grass is green", we would like to include this sentence, even though "color" is never mentioned there. Finally, the highest scoring segments are parsed to extract the information needed to answer the query. This is a costly process, hence the need to first weed out documents and even parts of documents, to focus on a few fragments. This final task is somewhat similar to IE, in that we are looking for some specific property, entity or relationship; the task is guided by the answer type (i.e. are we looking for a person's name, or a number?).

QA systems usually have to perform several tasks:

- understanding the question, including deciding the *focus* of the question (exactly what is being asked) and the *type of answer* that the question expects. The most common types are *list questions*, which require a list of simple answers (for instance, "Name 15 religious cults", where the list is further quantified); *punctual* or *factoid questions*, which require a simple data piece (for instance, "What is the height of Mt. Everest?"); *comparative questions*, which require a comparison among data pieces (for instance, "What is the cheapest car sold in the U.S.?"); *concept* or *definition ques-*

tions, which ask for the definition of a concept (for instance, "What is an antigen?"). Others are *why* questions (which involve finding the reason behind some event/fact), *hypothetical questions* (what would happen in a what-if scenario), and *cross-lingual questions* (i.e. translations).

- retrieving answer parts, and putting them together. The answer may not be directly found in any document in the collection; the system must try to find relevant documents, where at least part of the answer may be; further search within the document to find the relevant segments (sentences, paragraphs), and then put together the answer from the information found.
- bridging the gap between the question and the information found, usually by supporting some reasoning capabilities. For instance, a question like "What is done with worn and outdated flags?" (example from [46]) is unlikely to have a *direct* answer in a document; therefore, relevant, related information has to be used to compose an answer.

Corresponding to this, most QA systems have an architecture with several discernible components:

1. a *question classifier module* that determines the type of question and the type of answer.
2. a *document retrieval module* that attempts to identify, quickly, which documents in the whole collection are likely to be useful (contain -part of-the answer).
3. a *filter/analysis module*, which does further processing on the documents selected by the previous module. Sometimes this module also marks a part of a document as interesting (so that the next module does not work on the whole document).
4. an *answer extraction module*, which works on the documents selected by the previous node by focusing on selected parts of the document and trying to extract an answer to the original question.

Most systems use simple, keyword-based techniques for step 2. This allows to quickly search through a large corpus. Then, progressively more complex techniques are applied to smaller subsets of data. Many systems use templates to deal with step 3. When this does not work, or the system tries to do more sophisticated analysis, Information Extraction (IE) techniques are used. In particular, *entity extraction, co-reference analysis*, and sometimes *link extraction* are attempted. Some of the questions, like *why* questions or hypothetical questions require this more in-depth type of analysis. Some systems do (limited) reasoning, like common sense reasoning or temporal reasoning, in order to come up with good answers.

However, the most important -and most distinctive aspect of QA, separating it from IE- is the analysis of the question. Different types of questions require different types of answers, and usually lead to different processing strategies. Hence, analyzing the question correctly is vital for good performance. One of the most important tasks is to decide what type of answer the question is asking for[1].

QA systems have attracted considerable attention from the research community lately. The TREC series of conferences, sponsored by NIST, are a very good sample of research in the area ([1, 2, 92]).

QA systems can be divided into

- *Closed-domain*, systems that deal with questions under a specific domain. This allows the system to exploit domain-specific knowledge.
- *Open-domain*, systems that do not have their domain constrained in advance. Such systems are harder to create, but usually have large collections of documents to rely on.

8.3 GQs in Natural Language Analysis

One of the best assets of the generalized quantifier approach is the large body of knowledge already accumulated in linguistics and logic research and that can be applied to the present task. It is well known that using generalized quantification in linguistic analysis has a long and rich tradition, from issues of expressive power ([91, 61, 62, 101, 97]), to dealing with anaphora ([36]) and plural and mass terms ([71]); what is interesting to point out is that there is also discussion of the semantics and pragmatics of questions in this line of research; see ([20, 39, 42, 96]) as a sample. Generalized quantifiers have already been applied to computational tasks, besides query languages ([82, 90]). Finally, we point out that application of logic methods to QA is a nascent area of interest ([75, 46]).

The seminal paper by Barwise and Cooper ([16]) made GQs known to an audience of linguists working on formalizing natural language. Earlier work by Montague had used some of the concepts implicitly, but [16] made the connection explicit and exploit it to show what a promising avenue of research it was. This was followed by an explosion of work in the area.

Linguists had always been dissatisfied with the ability of FOL to formalize natural language phrases. Leaving aside the limitations of FOL, even for simple phrases there is something lacking when two similar sentences like (1) and (2) below have to be formalized in such a different way:

(1) Every student is smart.

is translated as

$$\forall x(Student(x) \rightarrow Smart(x))$$

using some obvious predicates, while

(2) Some students are smart.

---

[1] A large typology can be found at
http://www.isi.edu/natural-language/projects/webclopedia/Taxonomy/taxonomy_toplevel.html.
Note that some types are motivated by the processing that follows.

is translated as

$$\exists x(Student(x) \land Smart(x))$$

If both sentences are so similar (only the determiner, *some* and *all*, and the verb number, change), why are the translations so dissimilar? Why is there no *uniform* correspondence between parts of the sentence (subject, main verb, object) and parts of the formula? [16] proposed the following approach: a *noun phrase*, which usually acts as the subject of a sentence, is a linguistic structure that consists of a *noun* plus some (optional) parts, like *adjectives* or (in the case of quantified sentences) *determiners*: every, most, some, at least five,... are all examples of determiners. [16] identified determiners with GQs of type $(1,1)$. The first set corresponds to the rest of the noun phrases, the second set to the verb phrase that makes up the rest of the sentence (verb plus object). Thus, (1) above is analyzed as (again, using obvious notation for sets and predicates)

$$every(\{x \mid Student(x)\}, \{y \mid Smart(y)\})$$

where *every* corresponds to the quantifier $\mathbf{all}^{(1,1)}$ that we are already familiar with. There are multiple advantages to this approach:

- To formalize (2), simply substitute *every* by *some*, which of course corresponds to quantifier *exists* of type $(1,1)$. Now both formulas have exactly the same form.
- Moreover, the NP (subject) can be given a denotation of its own, by turning the $(1,1)$ relation into a function that assigns, to each subset of the domain, a set of subsets of the domain.
- The approach can be extended easily to deal with other determiners. Note that some like *at least five* are *FOL*-definable, while others like *most*, are not.
- The approach can be extended easily to deal with quantification in other positions, and with multiple quantification or nested quantification, as in the sentence "each librarian read three book reviews" (note that the sentence is ambiguous. Much more on this later).

As stated, this approach has developed in a very fruitful line of investigation. The questions that appear here are somewhat different as before, because we are in a linguistic, rather than logical, environment. Thus, while each quantified NP can be seen as corresponding to a GQ, the converse is not true: that is, it seems that not all GQs have an expression in natural languages. Moreover, the GQs that *are* realized in natural languages seem to have some special properties. Most work has centered around identifying those GQs that are realized, their properties, and the kind of logics they give rise to.

To see this, we must note that, logically speaking, there are many quantifiers. For instance, in a universe with two elements, there are 2^{10} quantifiers of type $(1,1)$. This is due to the fact that is explained in chapter 6. However, of these only a few have a counterpart in natural language. The question is then, which ones and why?

Remark On many papers in the linguistic literature, quantifiers are presented as *functionals*, that is, the underlying n-ary relation is formalized as a function of $n-1$ arguments into the powerset of the domain. For instance, the quantifier **all**$^{(1,1)}$ is presented as

$$\mathbf{every}(A) = \{B \subseteq M \mid A \subseteq B\}$$

Quantifiers of type (1) are presented a functions from the powerset of the domain into the set $\{0, 1\}$. Clearly, this is just a notational issue, brought about because of the interpretation of syntactic types (like D, N, NP, VP) as functions that combine with others to generate more types (for instance, D takes an argument of type N to produce an NP). Here we stick with the relational notation to be consistent with the rest of the book.

We have already seen in chapter 6 that when formalizing natural language statements in a logic with generalized quantifiers, one has to be careful with the formalization as issues of ambiguity and interpretation need to be dealt with. The basic assumption in the formalization is that GQs correspond to natural language determiners (where determiners (D) are defined as categories that take a Noun (N) category to form a Noun Phrase (NP) category). Thus, the GQ corresponds to D, and a first argument corresponds to the N; together they form an NP. To complete the sentence, a second argument corresponds to a Verb Phrase (VP). Together with the NP, this formalizes the sentence. It is for this reason that quantifiers of type $(1, 1)$ are seen as *naturally* occurring in natural languages. Note that this creates some GQs that may seem strange from a purely formal point of view; for instance, the determiner *a* is captured by **some**$^{(1,1)}$, since *a student is working* is interpreted as *some student is working*; while the determiner *the* is interpreted as

$$\mathbf{the}^{(1,1)} = \{A, B \subseteq M \mid A \subseteq B \wedge |A| = 1\}$$

since *the student is working* is interpreted as *there is a set of students A with exactly one member; and all the students in A are working*[2].

Note also that this does not imply that other types of quantifiers cannot occur (in chapter 6 we used type (1) quantifiers); but any such occurrence should be examined carefully to make sure that it corresponds to a faithful translation. For instance, determiners like *nothing, something*, etc. can be considered as having a "dummy" first argument (the set of all things in the universe) or can be formalizes as having type (1). The same can be said of

---

[2] In fact, some analysis interpret sentences as *John walks* with the proper noun *John* denoting a set of sets (any set that includes the element denoted by the constant *John*); then the sentence can be formalized as others: **John**$\{x \mid walks(x)\}$. This makes analysis of similar sentences uniform, but note that this type of quantifiers are not generic!

all in sentences like *all cheered*[3]. And, due to the extremely flexible nature of natural language, one can find yet other determiners that do not quite fit into this type of analysis. We present a few examples:

- *only* can be analyzed as determiner (quantifier), as in *only students are smart*, interpreted by

$$\mathbf{only}^{(1,1)} = \{A, B \subseteq M \mid B \subseteq A\}$$

  However, interpreted this way **only** does not obey CONSERV, therefore prompting many linguists to look for alternative interpretations of **only**[4].

- Consider the sentence *more men than women play tennis*. To consider *more* as a determiner, it would seem that the most appropriate formalization would be a type $(1,1,1)$ quantifier, as in

$$more - then(\{x \mid Man(x)\}, \{y \mid Woman(y)\}, \{z \mid Play - tennis(z)\}$$

  defined by

$$\mathbf{more - than}^{(\mathbf{1,1,1})} = \{A, B, C \subseteq M \mid |A \cap C| > |B \cap C|\}$$

  The same analysis can be used in sentences with *fewer men than woman* ... or *as many men as women*.... Interestingly, the notion of CONSERV can be extended to such types (types $(1, \ldots, 1, k)$ for $k \geq 1$ and $m + 1$ arguments, $m \geq 0$.

- Some determiners like *many, a few,...* are ambiguous and/or context dependent. Hence, it is unclear how to formalize them. They can still be considered as type $(1,1)$ provided that an agreement is reached about their intended meaning or the domain is equipped with some *measure* that can be used to determine the exact interpretation of the determiner.

In spite of these examples, there is agreement among linguists that most determiners can be seen as type $(1,1)$ quantifiers, and these are the most frequently studied in linguistics. For these reasons, from now on we restrict our discussion for the rest of the chapter to type $(1,1)$ quantifiers.

All GQs are assumed to follow EXT, since all determiners seem to have denotations that are independent of the context of quantification. The next most cited axiom is CONSER, which we already saw in chapter 3. We repeat the definition here in a simplified form:

Definition 8.1. *A GQ Q of type* $(1,1)$ *has CONSERV if* $Q_M(A, B)$ *iff* $Q_M(A, A \cap B)$.

---

[3] This example is from [103].

[4] Indeed, several papers have been almost exclusively dedicated to the analysis of *only*.

This condition is imposed because the first argument to a GQ is the rest of the NP that acts as subject of the sentence, and it plays a different role than the VP that is represented by the second argument. In effect, the NP restricts the interpretation of the VP. For instance, in examples (1) and (2) above it is clear that the property of being smart is predicated of a subset of the set of students. It is trivially true also that if every student is smart, then every student is a student and smart (and the same happens with *some*). On the other hand, this rules out quantifiers like the Hartig quantifier and the Rescher quantifier. This idea of restricting to a subset of the universe corresponds to the logical idea of *relativization*. In fact, we saw in chapter 3) that a type $(1, 1)$ quantifier Q follows EXT and CONSERV if and only if there is a type (1) quantifier Q' such that Q is the relativization of Q'. However, linguists do not use type (1) quantifiers -Westerståhl has pointed out that even determiners like *everything* and *nothing* can be handled with type $(1, 1)$ quantifiers by using a universal predicate to stand for *thing*. The reason is that many such quantifiers do not follow EXT, and CONSERV cannot be applied to them. This linguistic intuition, that a limited universe of discourse is always present (implicit or explicit) in natural language expressions, supports our claim in chapter 4 that quantifiers of type $(1, 1)$ are the appropriate ones for a query language.

A quantifier is trivial on universe M if it is empty or the universal relation on relations of M (of the right type). In linguistics, attention is usually restricted to non-trivial quantifiers. BEING A SIEVE.

Finally, this area of research answers an old question in formal semantics: Is FOL strong enough to formalize natural language? While intuitively the answer is clearly "no", giving a *proof* crucially depends on characterizing formalizations of natural language sentences and then determining the properties of said formalization. The corresponding formalization of sentences with expressions like *most* use the logic $L(most)$, just like the formalization of sentences like *half* or *more than half* use the logic with the corresponding generalized quantifier -and there is no way around this, since we know that the GQ *most* (or the GQ **half** or **more-than-half**) are not expressible in FOL. But it follows, then, that the logic $L(most)$ is strictly more powerful than FOL (as we have seen in chapter 3). Thus, this work has formally established what was long suspected.

8.3.1 Combining Quantifiers

When two or more quantifiers are used in a formula, forming a quantifier prefix, they can be combined to generate a new quantifier. This process is called *iteration* in [103]. For example, $\forall x \exists y \varphi(x, y)$ is written as the new quantifier $Q_{\forall, \exists}$ of type (2) defined by

$Q_{\forall, \exists} x, y \varphi(x, y)$ iff $\forall x \exists y \varphi(x, y)$

This can be generalized to an arbitrary number of type (1) quantifiers. Note that this generates a linear prefix. The idea is used to analyze complex sentences like

"Three boys gave more roses than dahlias to every girl"

where quantifiers **three** (of type $(1,1)$), **more-than** (of type $(1,1,1)$) and **every** (of type $(1,1)$) are combined into a quantifier of type $(1,1,1,1,3)$.

As another example, a sentences like "Most critics reviewed two books", which can be expressed by the QLGQ sentence

$$\textbf{most}^1\{x \mid \textbf{two}^1\{y \mid \text{Review}(x, y)\}\}$$

can also be expressed with iteration: one can define a new quantifier **most**^{1} $\circ$ **two**^{1} of type $(1, 1)$ as the composition of the two quantifiers above. Note that there is another linear reading of the example sentence, in which the same two parts are involved. This reading is expressed by permuting the quantifiers order (and hence reversing the scope) and creates another new quantifier.

Definition 8.2. *If $Q_1, \ldots, Q_k$ are type (1) quantifiers, their iteration $(Q_1, \ldots, Q_k)$ of type $(1, \ldots, 1, k)$ of $k + 1$ arguments is defined inductively by*

1. *$(Q) = Q$*
2. *$(Q_0, Q_1, \ldots, Q_k) = (Q_0, (Q_1, \ldots, Q_k))$.*

Lemma 8.3. *If all of $Q_1, \ldots, Q_k$ are CONSERV (EXT), then so is their iteration $(Q_1, \ldots, Q_k)$.*

Definition 8.4. *Given Q of type $(1, k)$, the negation of Q ($\neg Q$) is defined as usual. The inner negation of Q ($Q\neg$) in domain M is defined as $Q\neg(A, R)$ iff $Q(A, M^k - R)$. The dual of Q (D^d) is defined as $Q^d = \neg(Q\neg) = (lnotQ)\neg$.*

If Q is CONSERV (EXT), so are its negation, inner negation and dual.

Lemma 8.5. *(Negation Lemma)*

1. *$(Q_1, \ldots, Q_k) = (Q_1, \ldots, Q_{i-1}, Q_I\neg, \neg Q_{i+1}, \ldots, Q_k)$*
2. *$\neg(Q_1, \ldots, Q_k) = (\neg Q_1, Q_2, \ldots, Q_k)$.*
3. *$(Q_1, \ldots, Q_k)\neg = (Q_1, \ldots, Q_{k-1}, Q_k\neg)$.*
4. *$(Q_1, \ldots, Q_k)^d = (Q_1^d, \ldots, Q_k^d)$*

These equalities are needed to justify the logical equality of sentences like the following pair: "All but two students read six or more plays" and "Exactly two students read less than six plays" ([59]).

A few definitions are needed for the next results.

Definition 8.6. *Two iterations $(Q_1, \ldots, Q_k)$ and $(Q_1', \ldots, Q_k')$ are similar if Q_i and Q_i' are of the same type ($1 \leq i \leq k$) and are balanced if for each i ($1 \leq i \leq k$), all $A_1, \ldots, A_m$, whenever $B = \emptyset$ or $B = M$, $Q_i(A_1, \ldots, A_m, B)$ iff $Q_i'(A_1, \ldots, A_m, B)$.*

Definition 8.7. *Two quantifiers Q and Q' of the same type $(k_1, \ldots, k_n)$ are equal on products if for all universes M, all relations $R_i \subseteq M_i^k$ such that $R_i = C_1 \times \ldots \times C_{k_i}$ (R_i is a Cartesian product) if $k_i \geq 2$, then $Q_M(R_1, \ldots, R_n)$ iff $Q'_M(R_1, \ldots, R_n)$.*

Lemma 8.8. *If two iterations are similar and equal on products, then they are equal.*

This lemma means that iterations are determined by Cartesian products. This can be strengthened with a few more assumptions:

Theorem 8.9. *If two iterations are similar, equal on products, balanced and non-trivial then they are identical (that is, $Q_i = Q'_i$ for all i).*

A consequence of this is the Linear Prefix Theorem of Keisler-Walcoe, which is the case where all quantifiers are either $\forall$ or $\exists$, so the above theorem can be considered a generalization of Keisler-Walcoe to general type (1) quantifiers prefixes.

[62] also studies iteration of quantifiers from a linguistic perspective, and studies several properties of iterations. For instance, if all quantifiers in an iteration are (upward, downward) monotone, then the quantifier that the iteration defines is also (upward, downward) monotone.

We note, though, that not all quantifiers can be expressed as iterations of simpler ones; for instance "every boy likes a different girl" and "every student criticized himself" are two examples from [59] of complex quantifiers that cannot be broken down into two (a combination of "every" and "different" for the first example and of "every" and "himself" for the second).

We close this section by connecting this research with work on non-linear prefixes introduced in chapter 6. In linguistics, it was debated early whether non-linear quantifiers are needed for representing natural language. Hintikka ([49]) was the first to argue so, but his examples were very controversial, and many linguists remained unconvinced. Barwise ([15]) gave examples that included generalized quantifiers (Hintikka's examples were restricted to $\forall$, $\exists$) and gathered more support. Nowadays, there is broad quorum that *cumulative readings* are indeed necessary ([88]); even though it is still argued whether pure *branching readings* are needed, most researchers seem to be of this opinion. Finally, [89] tries to give conditions under which to distinguish cumulative and branching readings of natural language, but they are not necessary and sufficient -the problem remains open.

8.4 QLGQ in QA

Queries usually are not the answer to "How" or "Why" or "What is" questions, although they could in principle be if the actions, instructions, reasons or definitions are considered as data and stored in a database[5]. In this case, the questions can be paraphrased as *List the different ways in which...* or *List the reasons why...* or *Find the definition of....* Thus, this difference seems to do more with the way information is organized (and the type of things allowed in the repository) than a difference between question and query. Hence, despite some differences, questions and queries are obviously related; they are simply different parts of a spectrum. It seems feasible, then, to develop a single language in which both can be posed.

The essential intuition is that interrogatives can be also seen as GQs by considering them as relations between sets: the set provided by the NP and VP in the question, and the answer[6]. Different types of interrogative quantifiers can be considered; quantifiers of type $(1,1)$ are called *unary interrogative quantifiers* in [42], and quantifiers of type $(1,1,1)$ are called *unary interrogative determiners*. Some examples of interrogative quantifiers are

$$
\begin{array}{ll}
\textbf{who} & \{X, Y \subseteq M \mid X \cap PERSON = Y\} \\
\textbf{what} & \{X, Y \subseteq M \mid X \cap (M - PERSON) = Y\} \\
\textbf{which}_n^Z & \{X, Y \subseteq M \mid X \cap Z = Y \ \wedge \ |Y| = n\} \\
\textbf{which ones}^Z & \{X, Y \subseteq M \mid X \cap Z = Y \ \wedge \ |Y| \geq 2\}
\end{array}
$$

Note that the last two are only defined on a restricted form, that is, it is considered that in proper usage they always make reference to some context. For context dependency, we note that context may be explicit (either in the same sentence, or, in a dialog, in previous sentences) or implicit. In database queries (where there are no dialogs), the context is always explicit (i.e. most queries are phrased with standard quantifiers, as we have seen). In QA, the context may be implicit, and it usually will be in systems that handle dialogs. Note also that **who** is restricted to the PERSON domain, while **what** is restricted to the complement of such domain; this kind of constraint is useful when trying to determine adequate answers[7].

We illustrate the approach with an example. The question "Who takes CHEM 101?" gets formalized as

$$
\textbf{who}(\{x \mid Take(x, CHEM101)\}, Y) \ = \ (\{x \mid Take(x, CHEM101)\} = Y)
$$

Thus, an answer is a set Y that is (extensionally) equal to $\{x \mid Take(x, CHEM101)\}$. Formulas in QLGQ can be easily extended to incorporate this type of formula. The most important change is the need to accommodate *set variables* in the

---

[5] Although they usually are not. As we argue later, this is more an issue of *representation* than one of querying, though.

[6] This basic intuition is presented in [42], in which this chapter is heavily indebted.

[7] Even though any QA system incorporates this knowledge about **who**- and **what**-questions, and many times more sophisticated types ([50]), often this knowledge is implicit in the reasoning or the code of the system. The salient feature of this approach is that it incorporates these (and other) types of constraints in a declarative and organized manner.

language. This is due to the fact that answers are sets (when the answer is a factoid rather than a list, we have a singleton). However, unlike the queries we have previously shown, we cannot determine beforehand how this set will be constructed; in the end, it will be a list of constants. Therefore we cannot express it as $\{x \mid \varphi(x)\}$ where φ is a given formula in the language. Thus, we are simply indicating what makes a complete and correct answer (more on this later). As another example, the question "Which 3 take CHEM101?" is analyzed as follows: first, this question only makes sense in a context, i.e. we pick 3 out of some set determined contextually (this is the role of Z). Assume we are talking about students, i.e. $Z = \{x \mid Student(x)\}$. Then the formalization of the question is

$$\textbf{Which}_3^{\{x \mid \textbf{Student(x)}\}}(\{y \mid Takes(y, CHEM101)\}, Y \wedge |Y| = 3) \; =$$

$$((\{x \mid Student(x)\} \cap \{y \mid Takes(y, CHEM101)\}) = Y \wedge |Y| = 3)$$

Thus, the answer to this question is a set with three elements that obeys the restriction imposed by the formula. Finally, we note that sometimes this formalization leaves out useful information: for instance, the question "What classes does John take?" is formalized as

$$\textbf{What}(\{x \mid Takes(John, x)\} \cap (M - PERSON), Y) \; =$$

$$((\{x \mid Takes(John, x)\} \cap (M - PERSON) = Y)$$

However, we know that the restriction to things $(M - PERSON)$ is too wide; if there is a $Class(x)$ predicate, we can restrict ourselves to it.

Some examples of interrogative determiners are

what	$\{Z, X, Y \subseteq M \mid Z \cap X = Y \wedge |X| = 1\}$
which$^{\textbf{W}}$	$\{Z, X, Y \subseteq M \mid (Z \cap W) \cap X = Y \wedge |X| = 1\}$
how many	$\{Z, X, \{n\} \subseteq M \mid |X \cap Y| = n\}$
whose	$\{Z, X, Y \subseteq M \mid Z \cap X = Y \wedge \forall y \in Y \exists p \in PERSON \; Owns(p, y)\}$

Note that **what** and **which** may have singular and plural readings; the correct reading can be inferred from the complete question: "What students attend the class?" is plural, and "What student won the contest?" is singular. The definition given above corresponds to the singular reading of both quantifiers; in the plural reading, the condition $|X| = 1$ is substituted by $|X| \geq 2$. Note also that **which** is considered a context dependent quantifier.

As an example, the question "What is the class taught by Paul?" gets formalized as

$$\textbf{What}(\{x \mid Class(x)\}, \{x \mid Teaches(x, Paul)\}, Y) \; =$$

$$((\{x \mid Class(x)\} \cap \{x \mid Teaches(x, Paul)\}) = Y)$$

Note that determiners, unlike quantifiers, take an explicit restriction as an argument. However, this extra argument may not be the context, or the whole context, for the question. "Which students take CHEM101?" is interpreted as follows: now we are explicitly mentioning a context (the set of students), but in order to use **Which** there has to be a more restrictive context that applies (otherwise we would say "What students take CHEM101?"). Assume we were talking about Linguistics students (so $W = \{x \mid Major(x, Linguistics)\}$). Then we obtain

$$\mathbf{Which}(\{x \mid Student(x)\}, \{y \mid Takes(y, CHEM101)\}, Y) =$$

$$((\{x \mid Student(x)\} \cap \{x \mid Major(x, Linguistics)\}) \cap \{y \mid Takes(y, CHEM101)\} = Y)$$

The approach is extensible to *modifiers*, under the intuition that modifier interrogatives like **when, where, how** can be treated as quantifiers. To achieve that, a *sorted* approach is required, i.e.. one in which the domain has been divided into sets or *sorts*. We have already introduced the idea of sorts when we used PERSON to denote a specific subset of the domain; extending this idea yields other domains like PLACE and TIME. If one admits higher-order sorts, it is possible to represent concepts like MANNER, CAUSE and REASON (see [42] for the details)[8].

The modifier interrogative quantifiers are defined as follows. Let S be the union of all sets of sorts, and assume we have a number s of sorts. For any relation R of arity n, we extend it to a relation R' of arity $n + s$, where an element of each sort is incorporated: thus, $< r_1, \ldots, r_n, a_1, \ldots, a_s > \in R'$ iff $< r_1, \ldots, r_n > \in R$ and $a_i \in S_i$, the i-th sort. This will allow us to incorporate into the behavior of the modifier the implicit sort argument.

Definition 8.10. *For* $p \in PLACE$, $t \in TIME$, $m \in MANNER$, $c \in CAUSE$, $r \in REASON$,

where $\{R', R, Y \subseteq M \mid \{p \mid < r_1, \ldots, r_n, a_1, \ldots, a_n > \in R' \wedge < r_1, \ldots, r_n > \in R\} = X\}$
when $\{R', R, Y \subseteq M \mid \{t \mid < r_1, \ldots, r_n, a_1, \ldots, a_n > \in R' \wedge < r_1, \ldots, r_n > \in R\} = X\}$
how $\{R', R, Y \subseteq M \mid \{m \mid < r_1, \ldots, r_n, a_1, \ldots, a_n > \in R' \wedge < r_1, \ldots, r_n > \in R\} = X\}$
why$_c$ $\{R', R, Y \subseteq M \mid \{c \mid < r_1, \ldots, r_n, a_1, \ldots, a_n > \in R' \wedge < r_1, \ldots, r_n > \in R\} = X\}$
why$_r$ $\{R', R, Y \subseteq M \mid \{r \mid < r_1, \ldots, r_n, a_1, \ldots, a_n > \in R' \wedge < r_1, \ldots, r_n > \in R\} = X\}$

The p is one of the a_i in the definition of **where**, *and likewise for the other definitions.*

Note that we distinguish two senses of **why**, one asking for cause and one for reason. Again, how to actually set those apart in a natural language question may require some analysis.

As an example, the question "When did John take CHEM101?" can only be answered if the information about who takes what is augmented with

---

[8] Obviously, such domains may have internal structure; we don't enter into these details in this paper, but we'll have something more to say later.

temporal information. Thus, from *Takes(x,y)* we obtain *Takes'(x,y,p,t,m,c,r)*, which is meant to represent the fact that x takes class y in location p at time t in the manner m for cause c and reason r. Then, the analysis of the sentence is

$$\mathbf{when}(\{p \mid Takes'(John, CHEM101, p, t, m, c, r) \wedge Takes(John, CHEM101)\} = X)$$

denotes the answer as the set X of time points when John takes CHEM101. The question "Why did John take CHEM101?" is analyzed as

$$\mathbf{why_c}(\{c \mid Takes'(John, CHEM101, p, t, m, c, r) \wedge Takes(John, CHEM101)\} = X)$$

when it refers to the cause of John taking this class, and as

$$\mathbf{why_r}(\{r \mid Takes'(John, CHEM101, p, t, m, c, r) \wedge Takes(John, CHEM101)\} = X)$$

when it refers to the reason that he is taking CHEM101.

The approach can be extended to modifier interrogative determiners like **in which, for what (reason), at what (time), in which (manner)**; the interested reader is referred to [42].

8.5 CQA, QA and GQs

In this section we tie the material of the previous chapter with the material presented here, as an example of how research in one area can help illuminate aspects of work in another.

An ability of the approach is to help determine the type of expected answer, an information that QA systems put to good use. The approach correctly predicts that **who** questions asks for persons, **what** questions for objects (non person entities), and **how many** for cardinal determiners of the type **exactly_n**. Clearly, this is only the beginning of the typology needed by a QA system, which requires quite a bit more precision ([50]), but recall that we do not analyze the internal structure of the query, so this is just an initial assessment that can guide further processing. Furthermore, the notion of answer can now be formally analyzed. It is assumed that an answer is always a set, and that a question has only one *correct and complete* answer. Thus, if the question "Which students have taken all classes offered by Peter?" refers to Paul and Mary, this means that the only correct and complete answer is the set { Paul, Mary}. This discards the set {Paul} as a good answer, as it is not complete (i.e.. it is a *partial answer*), or the set {John} as it is not correct. This correct and total answer is the one that a QA system strives to find, but if it is not found, partial answers may be offered. Note that, for a given answer S, the set of all partial answers of S forms a lattice (the empty answer is excluded), and its elements can be given a partial order. Note also that this order can be used to *rank* partial answers, and to order questions[9]. For question f, let as denote by $Ans(f)$ the complete and correct answer to f. As in [42], we say that question f subsumes question g (in symbols, $f \leq g$) iff $Ans(g) \subseteq Ans(f)$. This relation is a partial order; if question f subsumes question g, then $Ans(g)$ is a partial answer for f. We can exploit this fact together with the following definition, which extends the one in the previous chapter:

Definition 8.11. *Let Q be a standard quantifier or interrogative quantifier. Then $Q(X)_1 = \{A \mid Q(A, X)\}$ and $Q(X)_2 = \{B \mid Q(X, B)\}$. Let D be an interrogative determiner. Then $D(X, Y)_2 = \{A \mid D(X, Y, A)\}$.*

Definition 8.12. *An interrogative quantifier Q is decreasing iff for all $X, Y \in M$, if $X \subseteq Y$ then $Q(X)_2 \leq Q(Y)_2$[10]. Similarly, an interrogative determiner D is decreasing iff for all $X, Y, Z \in M$, if $X \subseteq Y$ then $D(X, Z)_2 \leq D(Y, Z)_2$.*

The interesting fact, proved in [42], is that argument interrogative quantifies are decreasing, as are argument interrogative determiners. Therefore, there is an interplay between different questions and their answers that can be exploited by a system: starting with query g, we can obtain query f such that $f \leq g$ by relaxing g. Answering f will yield an answer that contains the answer of the original query. This ability can be used in several ways; for instance, queries may be relaxed to retrieve related, relevant information, in order to give the most informative answer. In QLGQ there are two ways to relax a query: the first one is to relax the GQ used, the second one is to relax the set terms. We give a brief example of each technique (this example is taken from [8]). For relaxing the GQ, assume the query *"Are all students enrolled?"*, expressed as follows in QLGQ:

$$\{() \mid \mathbf{all}(\{x \mid \text{Student}(x)\}, \{x \mid \text{Enrolled}(x)\})\}$$

The quantifier **all** can be computed as follows: $\mathbf{all}(X, Y)$ is true if $|X \cap Y| = |X|$. Suppose that in the course of computing the answer, we find out that actually $|X \cap Y| = |X|/2$. Then the answer is *no*, but we also know from our computations that half of them are (since $|X \cap Y| = |X|/2$ corresponds to the quantifier *half of Xs are Ys*), so we can answer *No, but half of them are.*

The second technique, relaxing the set terms, has been exploited before in the research literature. The novelty here is that we can determine when

---

[9] Note that there is also in QA a notion of exact answer, but it is somewhat different. Since in QA systems an important issue is that the paragraph returned may contain extraneous information, i.e. *more* than the answer, we have a possibility not considered here -such paragraphs would be considered completely incorrect in the GQ framework. Curiously, TREC 2002 judged such answers as inexact and did not allow them to contribute to the score ([99]).

[10] Note that for an interrogative quantifier, its second argument (denoted by $Q(X)_2$ is unique and is the answer.

it is a good idea to do so, based solely on the structure of the query. Since the answer depends not only on the set terms involved but on the quantifiers used, we have to determine what the effect of relaxing the set term is going to be on the query. This can be achieved of *monotonicity*, which was already used in the previous chapter to relax query answers.

This idea can be used in questions too, and may be useful for QA. Assume that a question asks "What are the mountains higher than 10,000 meters in Nepal?". No answer can be found to this query. Analyzing its formalization

$$\textbf{What } (\{x \mid Mountain(x) \land \{Height(x,y) \land y > 10,000\},$$
$$\{x \mid Mountain(x) \land Located(x, Nepal)\}, Y)$$

we can see that the set $\{x \mid Mountain(x) \land Height(x,y) \land y > 10,000\}$ is empty, i.e. there are no mountains on Earth (and therefore, in Nepal) which are higher than 10,000 meters. This could be used to relax the query by changing the constraint $y > 10,000$ to some realistic value (perhaps guided by already extracted information). Note that sometimes, no set term is empty, and yet there is no answer. If one asks "What are the names of mountains in Spain that are higher than 8,000 meters?" the answer is empty. One can analyze this as:

$$\textbf{What } (\{x \mid Mountain(x) \land Located(Spain, x)\},$$
$$\{x \mid Mountain(x) \land Height(x,y) \land y > 8,000\}, Y)$$

In this case, neither set is empty but the intersection is. This means that there are mountains in Spain, and there are mountains that high, but none of them is in Spain. Thus, we can choose to relax either set (or both) until an answer is obtained.

8.6 Challenges

One of the reasons GQs are so appealing in practical query languages is that they are high-level, declarative operators that can work with any type of elements, be they tuples in a relational database, XML fragments or chunks of natural language text, as far as set operators over such collections can be defined.

Since one of the directions in which research in databases is currently moving is that of providing a *unifying framework* for accessing all types of data, GQs can play a natural role in defining query languages that are *independent of the data model being used* and hence can work with *several* data models at once.

In this sense, the aim of the research presented here is to provide a unifying framework in which both QA and traditional data querying can be incorporated. By extending QLGQ syntax and semantics to work with these formalizations of natural language quantifiers, determiners and interrogatives, we have a common language that can be applied to both databases and collections of facts extracted from documents, thus providing a framework in which to integrate questions and queries. Moreover, the framework is declarative, high-level and extensible. For instance, the approach can be extended to multiple questions ("Which country was visited by which person?", or "Who bought what?") or to questions that utilize (declarative) quantifiers ("Who bought every item in the store?" or "Who visited at least two countries?"). Thus, QLGQ (with the extension proposed here) is very well suited to be the *common language* in which to represent both questions and queries that our framework calls for. Its similarity to the natural language (surface) structure means that the formalization effort is simplified, while the fact that it is a formal language with well defined semantics means that reasoning and search procedures can be carefully constructed and analyzed for correctness. Thus, it is a reasonable effort to formalize questions in QLGQ, while it is also a reasonable effort to represent (SQL) queries in QLGQ.

However, it could be argued that the framework presented is too general to be of real use in QA. Analyzing the task as a relationship between questions and answers misses much of the detailed work needed to produce answers. For instance, we have mapped questions to flat structures made up of predicate expressions; for quantifiers, the question was mapped to a single set expression/formula, while for determiners we divided the question into two parts -approximately corresponding to an NP/VP analysis of the English expression. However, real QA systems would strive to distinguish a *pattern* and a *focus* in a question, besides a set of keywords (possibly connected in a predicate-argument structure) ([87]). Thus, a ternary relationship between a question pattern, a question focus and an answer may be a more realistic framework. But the analysis may not stop there; a 5-ary relationship between a question pattern, a question focus, a question type, a question vocabulary (the keywords used) and an answer may be proposed. Different systems may propose and use different components. Many resources (like Named Entity Recognizers, thesauri or ontologies, semantic parsers) and many steps may be involved in generating such components. Our analysis deliberately stopped at the topmost (most coarse) level in order to abstract from these details. This is in line with out goal of providing a general framework, and is consistent with a need to further elaborate the framework. We next show some characteristics of the approach that show its promise, and discuss some of its shortcomings, in order to provide a better perspective.

Another important issue is that we have assumed all along that the right predicates are available to form the correct set terms for each phrase in the query or question. However, a large part of the work in a QA system is precisely to decide the right structures to represent the information in a query or a document (and therefore, the *vocabulary* available); thus, our assumption hides quite a considerable amount of work. But considerable research in QA is devoted already to the task, and can be reused in this framework.

A especially difficult problem to integrate QA and DQ is the different nature of information in databases and documents. In databases, the vocabulary comes given by the *schema* of the database (i.e., in relational databases, the name of the relations in the database and the names of the attributes on those relations). This schema is fixed beforehand, and is determined in part by issues of database design ([6]). In documents, on the other hand, the vocabulary is built dynamically based on what text analysis finds in the documents that constitute the collection. Thus, this vocabulary is not fixed in advance and comes determined by what the system is able to find -and, most of the time, natural language sentences yield structures that are considerably different from a database schema. Not only do sentences provide facts that do not adjust to a predefined schema; natural language freely mixes raw facts, metadata about the facts (including links among them), and information that is usually not captured in databases, like explanations, justifications, etc. Proof of this fact is that database query languages rarely -if ever- are used to ask *Why...?* or *How...?* questions. This is an extremely important and difficult issue, which is outside of the scope of this paper.

At the language level, there are also important problems that are to be resolved. For instance, even though factoids and lists can be dealt with in the proposed framework, definitions are not addressed. Even within the narrow issue of factoids, there are important differences between the way information is access in QA and in DQ. One important difference is that in a query, the type of answer is fixed by the query expression and trivial to determine; therefore there is no need for further analysis of the query expression. However, an interesting possibility that our approach opens up is the analysis of the query to determine if a *topic* and a *focus* can be associated with it, perhaps with the help of additional structures, like a thesaurus. This would allow us to reformulate a query over a database in order to obtain additional information from a set of documents.

One problematic case is that of superlatives. Assume the question: "Which is the highest mountain in the world?". In a database, we would expect to find a table with information about mountains; for simplicity, assume a table named Mountains with attributes name and height. To obtain an answer to the question, we first need to calculate the largest height, then look for it; this is broken down into two steps in SQL as follows:

```
SELECT name FROM Mountains
WHERE height = (SELECT max(height) FROM Mountain)
```

There is an embedded subquery (after the equal sign) that computes the maximum height using an *aggregate function*[11]; once this is done, the actual value returned is used to compute the overall query. This approach is not feasible in a QA system, for two reasons: first, it is much more efficient to look for a passage that directly answers the question (with a statement like "Mt. Everest is the highest mountain in the world"), instead of finding all mountain height information and doing a comparison. But second, even if we decided to carry out this approach (perhaps because a direct answer was not found), the database uses the *closed world assumption*, which means that it assumes that it has information about all the mountains in the world (at least, all the mountains that matter: if a mountain is not in the database, it does not exist) ([6]). This assumption makes the comparison sensible. However, texts are more open-ended; not such assumption can reasonably be made. Thus, QA systems will attempt to find a direct answer to the question by locating a sentence or paragraph in a document that states the needed information. Even though QLGQ as defined does not support aggregate functions, they are not difficult to incorporate into the language. The problem is that one should be able to regard such functions not only as a computation (as in SQL) but also as defining a complex term that may have an exact match in some frame or similar structure extracted from the document repository. Note that being able to consider such functions both ways opens up the possibility of obtaining an answer in two ways: if a QA system was unable to find a straight answer, it could try to gather information about mountains and their heights; compute the highest height among those found and use this information to return an answer -with the caveat that the answer was found indirectly and based on the available information. Note that if a document states "Mount Everest is the highest mountain in the world" we can be certain that we have a correct and complete answer (as far as the document is not mistaken), while if the answer is computed we do not have that certainty -but a computed answer, good to the best of our knowledge, is better than nothing.

Finally, we point out that some of the definitions given may have to be amended to deal with the difficulties (faced by all QA system) of gathering and extracting information from natural language documents. For instance, the completion needed to answer modifier questions may not always be available, so we need to relax it: instead of having a complete extension for each predicate used, we may have to define a set of extensions determined by whatever information is available: thus, given relation $R(r_1, \ldots, r_n)$, $R_p(r_1, \ldots, r_n, p)$ is an extension when only location information is available; $R_t(r_1, \ldots, r_n, t)$ is an extension when only temporal information is available, and so on. This is motivated by the fact that information contained in natural language sentences (or paragraphs) may vary widely, and partial information is more the

---

[11] Aggregate functions in SQL are functions that compute a value out of a set of values; currently, *minimum, maximum, count, sum* and *average* are recognized by the standard -although many others are to become also part of it.

norm than the exception. Also, spatio-temporal information is many times implicit and hard to determine. Note, though, that such limited extensions are still good enough to answer some questions (but not others), and can be used in our framework.

9

Extensions

So far, we have presented a basic language which is not all that dissimilar from other, existing query languages. The main *reason d'etre* for $QLGQ$ was to allow us to incorporate Generalized Quantifiers into the query language in an easy, natural way.

However, in order to allow for efficient processing, we have constructed a limited environment, and it is natural to ask if this environment could be extended, and if so, in which ways. Here we discuss several extensions to the language that seem of special interest, each one for its own reason. Since $QLGQ$ is so close to Datalog, adding adding recursion (fixpoint) to the language is an obvious step to take. Adding aggregates to the language is also an obvious step for a language that wants to have practical uses. Another, less obvious step is to allow for second order variables; we also explore this idea because of its connections to tasks like data mining. An extension to distributed environments is worthwhile studying, as the issue of quantification in distributed databases has barely been studied. Finally, an extension to other data models besides relational emphasizes that the concept of Generalized Quantifier is high-level and data model independent.

9.1 Datalog-like Languages

It is clear that $QLGQ$ has strong similarities with Datalog. The basic formulas are equivalent; it is only the set formation operator that makes a difference. Adding sets to Datalog, however, has already been proposed (REF). Thus, the only real differences are the presence of generalized quantifiers in $QLGQ$ and the ability to do *fixpoint* computations in Datalog.

9.1.1 Aggregates

To add aggregates to the language, we incorporate the following rule:

A.Badia, *Quantifiers in Action: Generalized Quantification in Query, Logical and Natural Languages*, Advances in Database Systems 37, DOI: 10.1007/978-0-387-09564-6_9,
© Springer Science+Business Media, LLC 2009

- if S is a set term, and F is an aggregate function (one of `min`, `max`, `sum`, `count`, `avg`), then $F(S)$ is a *term*.

This means that the result of an aggregation can be used in further comparisons. However, this term is different from others in that it has no name; hence, we extend the language with the *tag* operator: if l is a variable name not used in S, then $l : F(S)$ simply gives the name l to the term $F(S)$. Note that this is a variable that obeys all scope and so on rules. As an example, the query "'give the average salary" becomes

$$\{A \mid A : avg(\{sal | Emp(\_, \_, sal, \_)\})\}$$

while the query "give the average salary per department" becomes

$$\{dept, A | A : avg(\{sal | Emp(\_, dep, sal, \_)\}\}$$

Note that aggregation introduces some issues that are brand new. First, aggregation generates values, i.e. creates values that may not be among the database values, while all other queries are always guaranteed to return values from the database. Note that this is not the same as the ability to invent arbitrary values, which is a powerful database primitive. Second, aggregate terms do not have a name. All other values are denoted through some variable, and hence can be named (in SQL, attribute names are used to denote sets of values, while in logic-based languages explicit variable names are used). Aggregate terms are derived from attributes, but are not attributes; hence, it is necessary to give such terms a name in order to use them later.

9.1.2 Fixpoint

Adding generalized quantifiers to Datalog could be done if a set formation operator is added to the language -something that, we just mentioned, has already been proposed. The real issue is the following: if the ability to do fixpoint computations is also added, we have to realize that all the operators that are not *monotone* on the Datalog sense (which corresponds to our concept of upward monotone) create a problem. In particular, downward monotone quantifiers (including **no**, which represents negation) can only be added under certain conditions. Indeed, it is well known that adding negation to Datalog can only be done in the form of *stratified* programs; otherwise, the semantics of negation interferes with the fixpoint computation.

An example will clarify the situation. Let the binary relation `Parent(x,y)` denote that x is a parent of y. To calculate the `Ancestor(x,y)` relation, intended to denote that x is an ancestor of y (note that this relation can be seen as the transitive closure of `Parent`), a typical Datalog program has this form:

```
Ancestor(X,Y) :- Parent(X,Y).
Ancestor(X,Y) :- Parent(X,Z), Ancestor(Z,Y).
```

A Datalog rule can be divided into a *head* (to the left of the :- symbol, which is a single predicate, and a *body* (to the right of the :- symbol, which is a list of predicates. The variables in the head must appear in the body, and are considered universally quantified. Variables appearing in the body but not the head are considered existentially quantified. In our example, the head is the same for both rules (`Ancestor(X,Y)`, while the bodies are different. The variables `X`, `Y` appear in both head and body, while the variable `Z` appears only in the body. The first rule simply copies into `Ancestor` the contents of `Parent`. In the second rule, the same predicate as in the head is used in the body. The computation for the head is done by assigning a value to `Ancestor` in the body and computing a new value for `Ancestor` in the head. When the computation does not produce a difference between the old and the new values for `Ancestor`, we stop: we have reached a *fixpoint* for `Ancestor`. In any regular Datalog program, reaching a fixpoint is guaranteed. Intuitively, this is due to two facts: starting with finite relations in the database, Datalog programs can only produce finite results; and the application of Datalog rules either keeps the extension of a predicate of adds tuples to it, but never decreases it. Since the result cannot keep on increasing forever (or we would have an infinite result), it follows that at some point the extension of the predicate remains the same. In our example, the relation `Parent` is a finite set of tuples stored in the database. Starting with this, `Ancestor` gets a copy of this finite set of tuples. The first application of the second rule, then, becomes a self join on `Parent` (since at that point `Ancestor` is a copy of Parent). However, the results of this self join are added to `Parent` (note that all the tuples already present in `Parent` remain there). Thus, at this point `Ancestor` contains parents and grandparents. The process repeats, adding a generation each time. At some point, we run out of generations to add (if the number of initial tuples was finite, then the number of generations ought to be finite too).

How can this be expressed in $QLGQ$? Since we have union, clearly what we need are set terms to express the first and second rule. But we immediately run into a problem: in Datalog, results are *named*. This is needed for fixpoint, since reusing the name of a head in the body of a rule is what starts the fixpoint computation. In $QLGQ$, in contrast, set terms are *anonymous*. For instance, the first rule can be translated trivially as

$$\{x, y \mid Parent(x, y)\}$$

But the result is nameless. This means that there is no way to translate the second rule. Hence, we add to $QLGQ$ the capability to name set terms, with a simple rule: Let L be a set of labels (symbols); if S is a set term, and l is a label, then $l : S$ is a *named* set term. Thus, we write

$$Ancestor : \{x, y \mid Parent(x, y)\}$$

for the first rule, and the second one becomes simply

$$Ancestor : \{x, y(z) \mid Parent(x, z) \land Ancestor(z, y)\}$$

Now the program can be modeled by unioning these two formulas.

What are the semantics of the language now? Intuitively, it seems that we could exactly the same semantics as Datalog. And indeed, for simple examples like this one such semantics would work just fine, and therefore we obtain a language equivalent to Datalog. However, when we combine this feature with generalized quantifiers, a problem arises.

As stated above, Datalog formulas do not admit grouping or negation. While both can be added to the language (and there have been several proposals to do so), both constructs create problems with the fixpoint computation. For instance, in the typical example

```
Bachelor(X) :- Man(X), not Married(X).
```

a bachelor is defined as man who is not married. The number of bachelors in our domain may go up when the number of man goes up, but it will go down when the number of married man goes up. That is, if we add an element to Married, we have to *subtract it* from Bachelor. While this behavior is well understood, it means that we cannot freely use negation in a fixpoint computation. Intuitively, if the predicate being computed does not grow or stay the same at each step (conversely, does shrink at any step), then we have no guarantee that we will reach a fixpoint and our computation will terminate. We borrow an example from Ramakrishnan and Gerhke ([83]) to explain the situation: let Assembly(X,Y,Z) be a relation that holds if part X has Z copies of subpart Y (for instance, Assembly(car,wheel,4)). Note that subparts may have parts in turn. Then a recursive program to find out the components (subparts of subparts) of a part can be written as

```
Components(X,Y) :- Assembly(X,Y,Z).
Components(X,Y) :- Assembly(X,W,Z), Components(W,Y).
```

Two extensions to Datalog are the ability to create sets and the ability to use negation. They both present problems for the fixpoint computation. Assume we want to classify parts into big and small, depending on whether they use 3 or more copies of some subpart. Then we can write

```
Big(X) :- Assembly(X,Y,Z), Z > 2, not Small(X).
Small(X) :- Assembly(X,Y,Z), not Big(X).
```

The problem with this program is that, depending on the order in which the individual rules are applied to an initial database, we may end up with different answers. This is not the case for relational operators (or, in fact, for any operators we have considered so far, including generalized quantifiers). The problem with sets is illustrated with this example: we want to compute the number of subparts for each part, so we use the rule

```
NumParts(X, sum(<Z>)) :- Assembly(X,Y,Z).
```

The <> is the *set formation* operator on Datalog extensions that allow such operation, and is analogous to grouping in SQL. In fact, the example above can be easily written in SQL with grouping and aggregation. The following program tries to compute the number of all components (subparts of subparts) for each part:

```
NumParts(X, sum(<Z>)) :- TotalParts(X,Y,Z).
TotalParts(X,Y,Z) :- Assembly(X,Y,Z).
TotalParts(X,Y,Z) :- Assembly(X,Y2,Z2), TotalParts(Y2,W2,Z3),
                     Z = Z2 * Z3.
```

We first compute (recursively) the number of parts, then add them up. Unfortunately, the program only works if computation of `NumParts` is done *after* the computation of `TotalParts` has reached its fixpoint. What is the connection with quantification? Quantifiers that are downward monotone present a similar problem: if they are true of a certain set, they are going to be true of any subset of that set -so, even if the set in question loses elements, the quantifier still holds true. Thus, if we want out fixpoint computation of $QLGQ$ to terminate, we need (in a similar vein to Datalog) to make sure that the set term being computed is not an argument to a downward monotone quantifier[1]. Note that **no** is considered downward monotone -as expected, since **no** plays the role of negation in $QLGQ$. Thus, the query above can now be written in $QLGQ$ as follows:

$$NumParts : \{x, A \mid A : sum(\{z \mid TotalParts(x, \_, z)\}$$
$$TotalParts : \{x, y, z \mid Assembly(x, y, z)\} \cup$$
$$TotalParts : \{x, y, z \mid Assembly(x, y_2, z_2) \wedge TotalParts(y_2, \_, z_3) \wedge z = z_2 \times z_3\}$$

This query is then interpreted as in Datalog.

We close this section by noting that quantifiers that are not monotone, like **exactly five**, may present the same problem. Only upward monotone quantifiers are safe for fixpoint; therefore, the restriction proposed above must be applied to all quantifiers except upward monotone ones.

9.1.3 Higher Order Variables

$QLGQ$ is essentially a first order language, in that all variables range over *individuals* in the domain. What would happen if we allowed higher order variables, that is, variables over *sets* of individuals? Assume, for instance, the query

$$\{A \mid \textbf{half } A \ \{\}$$

where the variable A is a variable over *sets*. Then we can understand this query to mean: list all the sets of individuals A such that the set is in the relation **half**

---

[1] Technically, the set term must be parametric to get us into trouble, since only such set terms contribute values to a query answer, so we can focus on such terms.

with the set of individuals that results from evaluating the set expression $\{\}$ (which is a regular set expression and can be evaluated with standard $QLGQ$ semantics). This extension of the language opens up intriguing possibilities, but it is important to realize that the extra expressive power comes with extra computational complexity. In essence, by adding second-order variables we have added (some of) the expressive power of second-order logic and with (some of) its complexity. Since Fagin's theorem tells use that second order logic captures NP (REF), we know that this extension to the language is going to be problematic. We give an example to make this point more concrete. Assume that $\varphi(\overline{x})$ is an arbitrary formula in $QLGQ$; then the set term

$$\{A \mid \mathbf{all}A \ \{\overline{x} \mid \varphi(\overline{x})\}\}$$

We know that $\mathbf{all}(X, Y)$ iff $X \subseteq Y$. Hence, this query asks to compute all the subsets of the set denoted by $\{\overline{x} \mid \varphi(\overline{x})\}$. This is in essence the *powerset* query: given a set, compute its powerset ([6]). This query is not expressible in first order logic. It is a high complexity query: if a set has n elements, then there are 2^n subsets. Thus, computing the powerset query is exponential on the data. Even if the formula φ is very easy to compute, the simple task of enumerating all the parts of the result (the subsets) will lead to high complexity, too high for all but toy examples (when n is extremely small).

Complexity is one of the reasons most query languages avoid *second order features*, that is, the ability to manipulate sets as individuals: with variables for them, and the ability to do whatever is done with individual variables. But if complexity poses such a problem, why consider these features at all? Because they bring extra expressive power which may come handy for some practical queries. Consider, for example, the syntax introduced in subsection 9.1.1 to deal with aggregates, and assume that we have a database of books, where each book is made up of a collection of chapters -to simplify, assume relation `Chapter(bid,cid)`, where `bid` is a book identifier and `cid` is a chapter identifier. Then, the query

$$\{bid, C \mid C : \{cid \mid Chapter(bid, cid)\}\}$$

would return a *nested relation*, one where one of the elements of a tuple is not a single value (a number, or a string), but a set ([6]). In effect, we are using the ability to name sets and taking the set as a whole (instead of applying an aggregate to it, which converts the whole set into a single value) and using it as a member of the tuple that is returned as part of the answer. The nested relational model is a very interesting extension of the relational model, but it comes with query languages that usually (and naturally) allow for the powerset query to be expressed -thus making those query languages too powerful. But because the nested relational model treats sets as *bona fide* elements, it is a natural for adding Generalized Quantifiers. The reason we did not use nested relations as our basic model is because of the very high complexity of the associated query languages.

This issue shows another limitation of the language studied (and hence, another possible extension). We have been dealing with sets -the parametric set terms of $QLGQ$ allowed for the formation of sets of objects depending on a parameter. This can be considered a weak or implicit form of nesting; however, not more than this was allowed. The language always returns a flat (that is, non nested) relation as answer. Now, consider the following example: a database modeling a library as a collection of books. The library wants to purchase new books only if they bring original content, but many textbooks (especially at the introductory level) cover exactly the same content. Given a book b, we have defined above the set of chapters in the book, $Ch(b)$. This can easily be extended to a collection of books, B, by defining $Ch(B) = \bigcup_{b \in B} Ch(b)$. We say that a book b is *covered* by a collection of books B (or that B covers b) if and only if $Ch(b) \subseteq Ch(B)$. The library will purchase a new book b if there isn't a collection of books B already in the library such that B covers b. We would like to write a query that checks, for a given book b, whether this is the case. We need two things: the ability to form sets of books, but also the ability to form *sets of sets* -so we can build the set of chapters for a collection of books. We cannot do so in $QLGQ$: a parametric set will give us the set of chapters for a single, given book. Note that coverage, since it is based on the subset relationships, could be expressed by the quantifier **all**. However, the quantifier must now compare a set with the elements of a set of sets. So, if a new textbook named "Introduction to Logic" is considered for acquisition, we would need to write something like

> **all**
> $\{c \mid Chapter(\text{"Introduction to Logic"}, c)\}$
> $\{c \mid Chapter(b, c)\}_{\{Book(b)\}}$

where the set term $\{c \mid Chapter(b, c)\}_{\{Book(b)\}}$ must be understood as: consider a book b (for convenience, we assume a database relation Book); get the set of all chapters of b; do so for an arbitrary set of bs. Clearly, to solve this query we would need to be able to do the powerset query. And, as already explained, this introduces too high a complexity into the language.

The idea of higher order variables, on the other hand, opens up another intriguing possibility: to have variables for *quantifiers*. That is, given $QLGQ$ formulas $\varphi(\overline{x})$ and $\psi(\overline{x})$, the $QLGQ$ query

$$\{Q \mid Q\{\overline{x} \mid \varphi(\overline{x})\}\{\overline{x} \mid \psi(\overline{x})\}\}$$

with Q a variable over quantifiers, can be read as stating: given the two set terms defined by the formulas, find out what the relationship between the two set terms is. This, in essence, is a *data mining* task ([44]). Indeed, generalized quantifiers have been proposed as a framework for data mining tasks ([43]). As a simple example, the parametric quantifier

$$AR_{m,n}(X, Y) = \{X, Y \subseteq M \mid \frac{|X \cap Y|}{|X|} > m \wedge |X \cup Y| > n\}$$

can be used to define *association rules*: the parameter m plays the role of *confidence* and the parameter n plays the role of *support*. The $QLGQ$ query above can be considered a generalization of this idea, in the sense that no particular constraint is required -and no parameters are given. That is, one could specify a quantifier (for instance, AR if one is interested in association rules) but not specify the parameters -so that the system must determine m and n from the data. Usually, there are several rules that will hold, and the semantics of the system could be changed so that only the results that maximize the parameters (or the top k such results) are returned. But one could also not specify the quantifier at all, and therefore let the system determine what exactly to look for. Thus, a system solving the above rule would in principle be free to consider any type of relationship between sets. However, a number of practical issues immediately arise. First, there can be many relations between the sets in consideration. Second, relations between sets that are downward monotone could be "discovered" at several levels. Clearly, for some such relations only the "highest" level is of interest (intuitively, we want to know the largest sets for which the relation holds). On the other hand, for upward monotone relations, knowing the smallest sets for which it holds is the interesting part. Thus, just as in traditional data mining, one must determine which solutions, among many possible, are the *interesting* ones.

Finally, we point out that a language where several types of quantification are allowed would offer large degrees of freedom to explore the data; however, such languages would be very difficult to implement. As an extreme example, assume that the query

$$\{Q, A, B \mid Q \ A \ B\}$$

where Q is a variable over quantifiers, and A, B are variables over sets, is allowed. Such query simply asks for all relationships between all pairs of sets of database elements. While the result may contain some very interesting relations, many of them are bound to be trivial and/or well known (a problem that many data mining methods have), and such result is bound to be impossible to compute efficiently. Restricting at least some elements of the query to be *grounded* (variable free) and restricting the kind of quantifiers under consideration may yield interesting query languages for data mining.

9.2 Distributed Quantification

Distributed systems today, whether peer-to-peer or mediated, only support SPJ queries. They are unable to support complex queries even in simple cases with homogeneous schemas. Our viewpoint is that this is due to the fact that, implicitly or explicitly, such complex queries involve (generalized) quantification (that is, from our perspective in this work, relations between sets); and in order to compute these complex queries it is necessary to establish what the universe of discourse (the sets in question) entails. This is, unfortunately not always easy on a distributed environment, as each database works with a Closed World Assumption and is therefore not aware of what others contain. In this section, we explain the problem from our perspective with an example, and then argue (using the same example) that $QLGQ$ can help solve the problem.

9.2.1 Quantification and Distributed Databases

Distributed databases usually employ a wrapper-mediator architecture, although federated databases (under the peer-to-peer banner) are making a comeback. In both cases, the underlying idea is the same: a set of several databases behave *as if* they were parts of one single, integrated database. The illusion is created by designing a system that will answer the user's question taking into account all information available in all the databases. When the databases are heterogeneous, very complex problems of information integration must be dealt with. In such cases, decomposing the user's query into queries that can be understood by each system, and putting all the answers back together are technically difficult tasks. However, even in simple, homogeneous environments it may be difficult to answer user's queries. In this chapter, we concentrate on such homogeneous cases, and show how relatively simple queries cannot be answer by current systems.

We define a *distributed database* as a set of m different databases or *nodes*, and denote a distributed database D as $\{N_1, \ldots, N_m\}$. We call $\{1, \ldots, n\}$ the set I of indices. We will assume that each node can communicate with each other (peer-to-peer case) or with a special node called a *mediator* (wrapper-mediator case). The goal of query processing in this context is to process queries as if they were run against D, even though D is only a virtual database. We assume all databases to be relational. Further we assume *horizontal partitioning*, that is, the same schema is present in all the nodes. For any relation $R \in sch(D)$, we denote the part of R at node N_i by R_i (note that some R_i may be empty). Thus, for any relation $R \in sch(D)$, $R = \cup_{i \in I} R_i$. The assumption of horizontal partitioning provides a very simple framework to introduce our ideas and disregard issues of heterogeneous information integration, which are outside our scope. Note that only the N_i are materialized, that is, only the R_i exist, the R being virtual. One important consequence of our assumption is that any relational algebra or SQL query q against D can be run, without any modification, against any node in D, since the schema is the same everywhere. We denote by $[\![(\!(\,]\!]_q)_D$ the result of evaluating q against the distributed database D, and $[\![(\!(\,]\!]_q)_{N_i}$ the result of evaluating q against node N_i.

We classify relational queries as follows: a query is SP if its Relational Algebra expression is made up of (perhaps repeated) applications of the Select and Project operators. Such queries can be readily expressed in SQL with a SELECT ... FROM ... WHERE statement, with no subqueries or grouping and a single relation in the FROM clause. A query is SPJ if its Relational Algebra expression is made up of (perhaps repeated) applications of the Select,

Project, and Join operators. Such queries can be readily expressed in SQL with a `SELECT ... FROM ... WHERE` statement, with no subqueries or grouping. Finally, a query is SPJ+ if its Relational Algebra expression necessitates set operations (union, intersection, difference) and/or grouping. Such queries are expressed in SQL in several ways. A *quantified query* is a SPJ or SPJ+ query.

For this chapter alone, we restrict basic formulas in $QLGQ$ to mention one relation name only. Thus, formulas like $R(x,y) \land S(y,z)$ are forbidden. However, such a query (which expresses a join) can always be written using the Generalized Quantifier **some**[2]. Thus, basic queries in this version of $QLGQ$ can only express SP queries; SPJ or SPJ+ queries require quantification.

The problem that we are dealing with can be stated, from our perspective, as follows: most systems proposed in the literature for distributed query processing are limited to handling SP queries (or some simple form of SPJ queries). This is due to two facts: one, most such systems deal with the (extremely complex) issue of integration of heterogeneous information. Two, many queries beyond SP cannot be computed in distributed environments by the standard procedures. To be more precise, if a user submits an SP queries q to a distributed database D that is horizontally partitioned, q can be answered by sending it (unchanged) to each node in D and then unioning the answers. In effect, we have that $sem(q)_D = \cup_{i \in I} sem(q)_{N_i}$. However, this is not true for SPJ or SPJ+ queries. Such queries can still be answered; the problem is that there is no known general procedure to handle them, and processing them as SP queries (that is, by sending copies of the query to each database and then combining the results) fails to deliver a correct answer. We next show with an example how this regular processing fails.

Imagine a large company which uses a series of a databases DB_1, DB_2, DB_3. The company is geographically distributed and so are its databases. The employees of this company are very mobile, and they work in projects at different sites. Each site keeps track of who is working on what. Relational technology is used to implement the databases. Assume each database has a relation *Works-in(ename,pname)*, where *ename* is the employee name and *pname* is the project name, but each database only sees its own part of the world and not all of it (i.e *works-in* can be considered as horizontally partitioned among databases). We assume a wrapper-mediator architecture ([105]). We use the term *system* to refer to the mediator and all the databases together. When the system receives a query, the mediator breaks it down as needed and sends each part to the database that contains information about it. Then answer is then put back together by the mediator and offered to the user.

Now consider the databases DB1, DB2 and DB3 with the following data:

---

[2] Note that, technically, queries that ask for a Cartesian product cannot be expressed in such a language. This can be solved by introducing the *trivial* GQ, one that is always true of any arguments.

DB1		DB2		DB3	
Works-on		Works-on		Works-on	
ename	pname	ename	pname	ename	pname
Smith	Star-Wars	Smith	ACME	Jones	Star-Wars
Jones	ACME	Jones	Thunderbold	Edward	ACME
Lewis	Star-Wars	Lewis	Star-Wars	Johnson	Thunderbold
Edwards	Star Wars				

and the following queries:

- Query 1: "Which projects does employee Smith work in?". This is an SP query that can be simply answered by sending the query to each database and taking the union of the answers. Clearly, the answer is the set {Star-Wars, ACME}.

- Query 2: List the people working on some project that Smith works on. This requires a self-join of $Works - on$ with itself. In Relational Algebra, we can write (with Works-on' a renaming of Works-on)

$$TEMP = \pi_{ename',pname'}(\sigma_{ename='Smith'}Works-on) \bowtie_{pname=pname'} Works-on'$$

and then the solution is simply $\pi_{ename'}(TEMP)$. If we send the above query to each database, DB1 will return the answer {Lewis, Edwards}, and DB2 and DB3 will return the empty answer. However, the real answer over the system is {Lewis, Edwards, Jones}. It can be seen that Jones and Smith are both working on ACME but since the tuples specifying so are in different databases, they are never part of any local join; therefore Jones is not retrieved. In SQL, the query could be written as

```
SELECT WK2.ename
FROM Works-on WK, Works-on WK2
WHERE WK.ename = 'Smith' and WK.pname = WK2.pname
```

If this query is run against each database separately, we obtain the same answers as before -incorrect ones. Note that this is an SPJ query.

- Query 3: List the employees who do not work on a project that Smith works on. This query can be formulated as the negation of the previous one, which in Relational Algebra is achieved with set difference: $Works - on - TEMP$, and then again projecting on the employee name. In SQL, the query could be written in a number of ways: using subqueries and NOT IN, correlated subqueries and NOT EXIST, or simply set difference as above. In either case, DB1 yields the answer {Jones}, DB2 yields the answer {Jones,Lewis} and DB3 yields the answer {Jones,Lewis,Johnson}. However, the correct answer is {Johnson}. DB1 has Smith working on Star-Wars only; because of CWA, it assumes that Jones is working on ACME only and that is its answer. DB2 has Smith working on ACME; because of CWA, it assumes that Jones and Lewis are working on Thunderbold or Star-Wars only (respectively), and that constitutes its answer. Finally,

DB3 does not have Smith working in anything; thus it returns all employees it knows about (Jones, Edwards and Johnson) as an answer. Again, no answer is correct. The correct answer (Johnson) appears only as part of DB3's answer.

- Query 4: List the employees who work on all the projects that Smith works on. This query can be formulated in Relation Algebra as

$$Works - on\ DIV\ \pi_{pname}(\sigma_{ename='Smith'}Works - on)$$

where DIV is the *division operator*. As it is well known, this operator is not really supported in the algebra, but it can be expressed with a combination of Cartesian product and negation. Thus, the real expression is much more complex. Likewise, expressing this query in SQL may take several (nested) subqueries. In the end, DB1 has Smith working on Star-Wars only; because of CWA, it assumes that Smith is not working on anything else, and therefore its answer is Edwards. DB2 has Smith working on ACME; because of CWA, it assumes that Smith is not working on anything else. However, as nobody else is working on ACME, it returns the empty answer. Finally, DB3 does not have Smith working in anything; vacuously, it will return all employees (Jones, Edwards, Johnson). Again, no answer is correct, as Jones is the only person working in all of Smith's projects (ACME, Star-Wars).

- Query 5: List the employees working on at least 2 projects that Smith works on. This query can be expressed as two self-joins of $Works - on$ with itself:

$$TEMP1 \bowtie_{ename=ename' \wedge pname \neq pname'} TEMP1'$$

in SQL, we would use a group-by and a count. In either case, in our example it can be seen that all databases would return the empty answer, missing Jones (him and Smith work on both ACME and Star-Wars).

Thus, none of the above five queries is answered correctly by the system. Each individual database uses CWA, which prevents the individual databases from realizing that they do not have complete information. Note that a mediator cannot challenge the individual answers, since they are correct as far as each individual database is concerned. But if the mediator tries to manipulate the individual answers to get the final (correct) answer, there does not seem to be any easy way to reconstructing the right answer. Note, for instance, that in Query 3 Jones is on all three answers, but it is not part of the answer, while in Query 5 all answers returned are empty.

Remark (for people with background on information integration) Clearly, the problem we are dealing with can be seen as a restriction on the general problem of integrating distributed information. This is an extremely hard problem because usually the integration process does not determine a database, but creates only partial knowledge (Libkin's paper on PODS'06). In our setting, however, we can claim that all STD (source-to-target) formulae are of the form

$$\phi_t(\overline{x}) : -\psi_s(\overline{y})$$

where $\overline{x} \subseteq \overline{y}$. This is due to the fact that we only use horizontal partitioning and therefore the schema of the final database is fixed. In this setting, all problems traditionally addressed by research in information integration are likely to be a) decidable and b) tractable (PTIME). However, from a practical point of view this is not enough. Our assumption is that some questions (queries with quantification) while doable in polynomial time (if our conjecture is right) do not have an easy expression in current settings (SQL). Hence, using GQs can be seen as an attempt to give efficient solution to a problem which, in principle, has one -but one that has not been developed yet.

9.2.2 Computing Distributed Quantification

In order to understand how $QLGQ$ helps compute answers in a distributed setting, we first sketch an evaluation procedure for $QLGQ$ in a unified environment. As stated in chapter 5, $QLGQ$ formulas can be written as a tree in a manner similar to relational algebra expressions, with basic formulas as leaves, and set terms or quantifiers as inner nodes, and evaluated *bottom-up*, by evaluating all basic formulas in subexpressions, and using the result of such an evaluation to evaluate inner nodes. The only rule is that before a node in the tree is evaluated, all its children need to be evaluated. Evaluation starts at the leaves, which are always basic formulas. Such formulas are evaluated against the database; this evaluation only (in our current setup) involves selections and projections. We then proceed up the tree. Non quantified formulas are simple SPJ expressions and are easy to evaluate. Quantifiers are then evaluated on their arguments (set terms); one way to do this is as indicated in chapter 5. The expression obtained at the root of the tree, which is always a (non-parametric) set node for a query, yields the answer to the query. Combining this strategy with the observations above, we can conclude that to process a $QLGQ$ query q in a bottom-up manner, we must process (at least) an SP query, and

1. nothing else, or
2. a quantified query.

Example 9.1. We use again the example of chapter 5 *"List the students who have at least 2 friends taught by Peter in some course"*, which was expressed in QLGQ as follows:

$$\{Sname \mid \textbf{at least two}$$
$$(\{fr \mid \text{Friend } (Sname, fr)\},$$
$$\{fr \mid \textbf{some}$$
$$(\{crs \mid \text{Lectures}(\textbf{Peter}, crs)\},$$
$$\{course \mid \text{Takes}(fr, course)\})\})\}$$

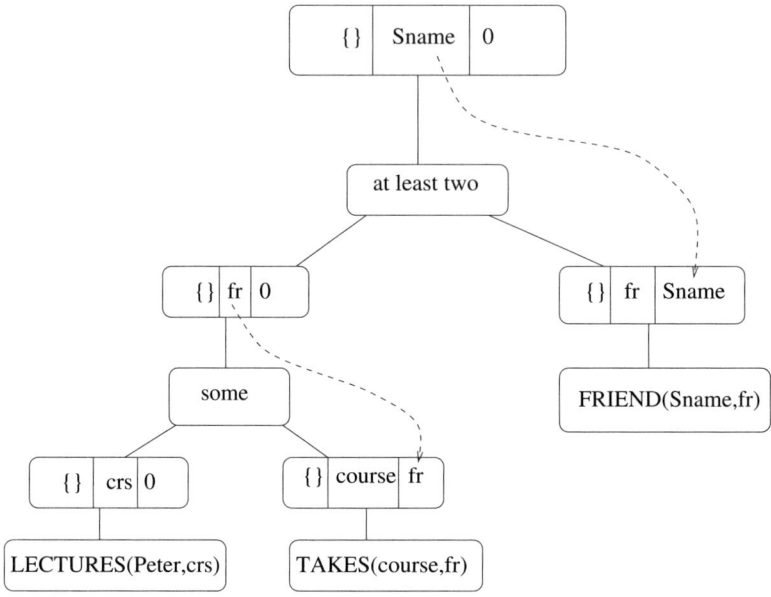

Fig. 9.1. Processing Tree

The tree for example 9.1 is shown in figure 9.1, with the nodes corresponding to set terms denoted by {,} followed by a list of bound variables and a list of free variables (0 is used to denote an empty list). Some observations help understand the figure:

1. Variables that are free at one level are bound on some level above (this relation is denoted by the dotted lines in the figure).
2. The union of the free and bound variable lists in a set node covers all free variables in the subtree of that set node.
3. The root node is always a set node with an empty free variable list.
4. Either the left or right subtree of a GQ node (or both) have non-empty free variable lists.

The computation can be seen as a tree traversal that starts at the leaves, finishes at the root, and proceeds in any order that respects the rule that all sons of a node must be evaluated before the parent node is evaluated[3]. Thus, either of $\{fr \mid \text{Friend}(Sname, fr)\}$, $\{crs \mid \text{Lectures}(\textbf{Peter}, crs)\}$, or $\{course \mid \text{Takes}(fr, course)\}$ can be evaluated first. When the latter two are

---

[3] This is similar to the rules of pure functional programming in which the only constraint is that before a function can be evaluated the arguments to the function must be evaluated.

evaluated, the quantifier **some** can be applied to the result of such evaluation. When this is done, and produces a result in turn, the quantifier **at least two** can be evaluated[4].

We assume that each relation R has been partitioned into *non-overlapping* subparts R_i each on node N_i. Call a query $q = \{\overline{x} \mid \varphi(\overline{x})\}$ *basic* if φ is a basic formula, and *quantified* if φ is a quantified formula. As stated above, all formulas in $QLGQ$ are either basic or quantified. As we saw, a basic formula is a relational formula perhaps with a comparison of terms, but in this context it cannot be anything else. Our basic approach can be outlined as follows: we assume the query is written in $QLGQ$. We evaluate the query with one of the strategies described below. The quantified formulas are of the type $Q(S_1, S_2)$ where S_1, S_2 are set expressions, and either S_1 or S_2 or both have parameters. If the set terms S_1 and S_2 are *simple*, then they correspond to an SP query, and they can be computed simply by sending a copy of the query to each node and unioning the results. Otherwise their definition involves some quantification. In this case, we proceed recursively. Thus, in the following we assume that both S_1 and S_2 are simple and concentrate on computing the quantifier.

There are several strategies to evaluate a $QLGQ$ query in this context. But we should note that the naive strategy (send a copy of the query to each node; union the answers) does not work. To see that, recall that the GQs we are dealing with are defined by (letting X and Y be the result of evaluating the set terms S_1 and S_2) $X - Y$, $Y - X$ and $X \cap Y$. However, it is clear that

- $X - Y \neq \bigcup_i X_i - Y_i$ (and similarly for $Y - X$).
- $X \cap Y \neq \bigcup_i X_i \cap Y_i$

Hence, we need some other strategy. Here we outline several possible approaches:

- *Query Conversion approach*: transform the $QLGQ$ query by using the interpreter of chapter 5. Then we are left with a SPJG (Select-Project-Join-GroupBy) query that can be processed in a distributed environment using standard methods. This approach may not be very efficient since distributed join processing is costly.
- *Materialized Sets approach*: compute X and Y by sending the subformula corresponding to S_1 (S_2) to each node and unioning the result (recall that we assume that these are SP queries). Then, in the mediator (or the peer that requested the result) compute the quantifier, either using the approach of chapter 5 or using set predicates. This approach is simple and correct. However, it is also clear that this approach may move a large amount of data.
- *Semi-materialized Sets approach*: note that while set difference and intersection cannot be completely distributed, the following holds:

---

[4] One of the consequences of this mode of operation is that $QLGQ$ queries can be, to a large extent, evaluated in parallel. We will not develop this issue any further.

- $X - Y = \bigcup_i X - Y_i = \bigcap_i X_i - Y$.
- $X \cap Y = \bigcup_i X_i \cap Y = \bigcup_i X \cap Y_i$.

Thus, we can follows a variant of the previous approach, where one of the two sets is materialized in the mediator or requesting peer; the materialized copy is then sent to each node, and the results returned to be unioned. This method can be considered a variation of the previous one.

- *Heuristic approach*: this is not so much another method as a collection of heuristics that can be tried before the other methods in order to improve performance. The heuristics are based on the fact that, in the approach of chapter 5, what we need are not the sets themselves (that is, $X - Y$ or $X \cap Y$) but their cardinalities. Then we can use several properties; for instance, $|A - B| \leq \Sigma_i |A_i - B_i|$ and $|A \cap B| \leq \Sigma_i |X_i - Y_i|$. That is, adding the local counts is sure to over-count the real global count. Then, for downward monotone quantifiers, we can act as follows: if the GQ is downward monotone and depends on $|A - B|$ that means its defining formula is of the form $|A - B| \leq n$ for some constant (or expression evaluating to the constant) n. Then if we do the local counts and adding them, and such counting are less than (or equal to) n, we know for sure that X and Y are in the relation that the GQ demands (since $|A - B| \leq \Sigma_i |A_i - B_i| \leq n$). The same is true for downward monotone quantifiers that are defined by $|A \cap B|$. If the quantifier is upward monotone, and depends on $|A - B|$ that means its defining formula is of the form $|A - B| \geq n$ for some constant (or expression evaluating to the constant) n. Then if we do the local counts and adding them, and such counting is less than (or equal to) n, we know for sure that X and Y are *not* in the relation that the GQ demands (since $|A - B| \leq \Sigma_i |A_i - B_i| \leq n$). The same is true for upward monotone quantifiers that are defined by $|A \cap B|$. Several variations of this idea are possible.

Applied to our previous example, this is what we get:

- Query 1: This is $\{Y \mid Works - on(Smith, Y)\}$ in $QLGQ$. This is simply an SP query and is processed as usual.
- Query 2: This is

$$\{Y \mid \mathbf{some}\{Z \mid Works - on(Y, Z)\}\{X \mid Works - on(Smith, X)\}$$

in $QLGQ$. In this case, there are two subformulas in the query (and two variables); the corresponding basic queries are $\{X \mid Works - on(Smith, X)\}$ (the set of all projects that Smith works on) and $Y : \{X \mid Works - on(Y, X)\}\}$, the set of all workers working on some project (i.e. a partition). Getting these sets evaluated is a simple SP query for each. In the first approach (Query Conversion), the whole query is translated into an SPJ query and evaluated as a standard relational query. In the second approach (Materialized Sets), the mediator gets these sets, and the quantifier **some** is computed by checking if intersection is empty. Note that the first set term is parametric, so we really check set intersection between the second set term and a partition of the first set term, created by partitioning the set term by Y. In the third (Semi-Materialized Approach), we materialize one of the sets only (say the second one, $\{X \mid Works - on(Smith, X)\}$), send a copy of this set to all the others, do partial checks there and send the results to the mediator. Again, on each node we compare the materialized term to a partition based on Y, so what we send to the mediator is the value of Y and the result of the corresponding test (say, True or False depending on whether intersection was not empty, or it was). The union of results here is the combination of all tuples with the same Y value into a single tuple (the combination is achieved with Boolean disjunction). In the fourth approach (Heuristics), we note that all we need is that $|X \cap Y| \geq 1$, so we proceed to compare local copies X_i and Y_i of X and Y at node N_i (again, note that X_i will actually be partitioned, and therefore several results will be obtained). Such local results are sent to the mediator, which will combine them (addition can be used to combine the counters). Now, if all values of the Y parameter are reported as passing the test, there is nothing else to do: each value found a local match, and this is all we need to know. Those values of the Y parameter that did not find a local match may have a match elsewhere, and they have to be checked using one of the previous techniques. However, only such values (without a local match) need to be checked, others do not need to.

- Query 3: This query is

$$\{Y \mid \mathbf{no}\{Z \mid Works - on(Y, Z)\}\{X \mid Works - on(Smith, X)\}$$

in $QLGQ$. Note that this query is the same as that of Query 2; the only difference is the quantifier. Thus, it can be computed in the same manner. In fact, the same is true of all other queries. Query 4 can be written as

$$\{Y \mid \mathbf{all}\{Z \mid Works - on(Y, Z)\}\{X \mid Works - on(Smith, X)\}$$

while Query 5 can be written as

$$\{Y \mid \mathbf{at\ least\ 2}\{Z \mid Works - on(Y, Z)\}\{X \mid Works - on(Smith, X)\}$$

We have basically decomposed each complex query in a series of SP queries which can be computed without any need for further reasoning. Clearly, there is room for improvement in the proposed approaches. However, as far as we are aware the standard approach is unable to cope with complex, quantifier queries at all.

9.3 Other Data Models

As stated earlier, quantification is independent of the data model used; therefore, it can be applied to other models besides the relational one. We sketched before on such possible extension, where the nested relational model was used. In this section, we present another extension to a different data model, that of *semistructured data* ([5]). The strategy is very simple: by changing the basic formulas in QLGQ to use XPath expressions, quantification is applied to semistructured (XML) data. Of course, other changes would have to be made to accommodate the power of XQuery or similar languages. The most significant one is that selectors in set terms should be allowed to build complex structure, in order to allow restructuring of a given database. Note, though, that if this is allowed, then quantification can be over complex objects. Thus, we need to make sure that types across set terms are *compatible*, in the sense that they can be meaningfully compared. Technically, we need to define an equality predicate on such complex terms. The idea is to exploit the fact that restricted semistructured models like XML are based on *trees*, and to declare two tree types compatible if and only if they are isomorphic (more loose definitions could use the idea of Bi-simulation, but that makes the definition of an equality predicate quite tricky). For isomorphic trees, then, equality can be defined recursively in a simple manner. Another significant challenge is that in XML we have *order*; such order may have to be taken into account in order to faithfully interpret the semantics of XQuery[5].

Using this idea, a language very similar to QLGQ can be used on an XML database. We give an intuition for this approach with an example. Assume an XML database like the following:

```
<catalog>
 <bookstore>
  <book>
   <title>Italian Cooking</title>
   <author>100</author>
   <year>2005</year>
   <price>30.00</price>
  </book>
  ...
 </bookstore>
 <authors>
  <author id='100'>
   <fname>Giada </fname>
   <lname> De Laurentis</lname>
  </author>
  ...
 </authors>
```

---

[5] One could argue that Generalized Quantifiers are set predicates, and hence should be interpreted without regards to order, even when one exists. The fact is, though, that some quantification is already allowed in XQuery, and it is introduced in an ordered context ([74]).

Then a query could be *Select the last name of authors such that all of their books are less than \$30*, which could be written as

> { *doc("books.xml")/authors/\$x/lname* |
>   **all**
>   {*book | doc("books.xml")/catalog/bookstore/book[author = \$x]*}
>   {*book | doc("books.xml")/catalog/bookstore/book[price < 30]*}

Note the need to specify here where the values (x) are coming from. This is due to the fact that we are dealing with complex types, and we can request values from different parts of the types in the database. Note also that, as before, one set term (the first one) is parametric, while another one (the second one) is not. This query would be written in XQuery as follows:

```
for $x in doc(''books.xml'')/catalog/authors
where every $y in doc(''books.xml'')/catalog/bookstore/book
      [author=$x] satisfies $y/price < 30
return $x/lname
```

Note that XQuery already possesses an **every** predicate; unlike SQL, both existential and universal quantification are present. However, optimization can be quite complex. Unnesting is possible ([74]). The situation could be improved by allowing generalized quantification in XQuery in a manner similar to SQL: subqueries are used to denote sets of (perhaps complex) elements. Once variables are introduced, a set term could be made parametric (dependant on the value of a given parameter). A generalized quantifier can be used as a filter in the **where** clause. As a simple example, the query above would look like this:

```
for $x in doc(''books.xml'')/catalog/authors
where every
      ($y in doc(''books.xml'')/catalog/bookstore/book[author=$x])
      ($y in doc(''books.xml'')/catalog/bookstore/book[price< 30])
return $x/lname
```

This query looks more complex than the original XQuery. This is due to the fact that the second set term can be specified by a simple predicate over an already bounded variable. This is always the case due to the syntax and semantics of XQuery: the second argument to a quantifier is a predicate. However, allowing a second subquery as argument means that arbitrary expressions can now be used in quantification. We retake our earlier example, where each book was made up of a collection of chapters. This would naturally be expressed in XML with a nested structure:

```
<catalog>
 <bookstore>
  <book>
```

```
<title>Italian Cooking</title>
<chapters>
 <chapter> Introduction </chapter>
 <chapter> aperitifs </chapter>
 <chapter> Entrees </chapter>
 <chapter> Desserts </chapter>
 <chapter> Drinks </chapter>
</chapters>
</book>
 . . .
```

Then the query *list pairs of books such that the chapters in one are contained in the chapters in the other* could be written as

```
for $x in doc(''books.xml'')/catalog/bookstore,
    $y in doc(''books.xml'')/catalog/bookstore,
where every
      ($z in doc(''books.xml'')/catalog/bookstore/$x/chapters)
      ($z in doc(''books.xml'')/catalog/bookstore/$y/chapters)
return ($x, $y)
```

Note that here we are comparing two separate sets, and it could be more efficient to process this query by generating (separately) the set of chapters for each book, and then comparing sets for inclusion. In the traditional processing, we would generate a set of chapters for a given book, and then "probe" the other books to check inclusion indirectly, by comparing chapter by chapter. Also, we observe that a query optimizer should be able to recognize the commonality on the path prefix of the expressions that are arguments to the quantifier, and take advantage of this fact. All in all, the first query above should be no harder to optimize or process than the original one, and a generalized quantifier-aware processor could do a better job with the second query than a traditional one. And, interestingly, in XQuery one can name the whole set using the **let** clause, where variables get bound to sequences (ordered sets), therefore imitating the mechanism introduced earlier of having second-order variables. Hence, we could write the second query above as

```
for $x in doc(''books.xml'')/catalog/bookstore,
    $y in doc(''books.xml'')/catalog/bookstore,
let $xchapters := doc(''books.xml'')/catalog/bookstore/$x/chapters;
let $ychapters := doc(''books.xml'')/catalog/bookstore/$y/chapters;
where every $xchapters $ychapters
return ($x, $y)
```

This is not currently allowed by XQuery, but we can see how it would work. Liberating the **every** operator to work on two arbitrary collections gives more flexibility when writing a query.

While one may argue that we do not seem to have gained much from these examples, the main advantage of Generalized Quantification, again, is its generality: under this approach, queries asking for authors such that half/most/ten percent/... of their books are under $30 can be written in exactly the same manner as the first query -simply change the quantifier used. And, once again, the effort to interpret the query -to produce the relevant counts, and use them as filters- is left where it should be, with the query engine.

10

Conclusion

In the previous chapters, we have explored the use of Generalized Quantification in practical applications. We have centered the work on query languages, but have progressively extended the scope to deal with questions and (limited) natural language analysis, as it is used in Question Answering. After introducing the basic logical concepts in chapter 2 and the concept of Generalized Quantifier in chapter 3, we presented the query language $QLGQ$ in chapter 4. This language was conceived basically as a vehicle to explore the use of generalized quantification, and is not very different from existing (logic-based) query languages, but it allow us to explore adding (and efficiently implementing) standard (type $(1,1)$) quantifiers to the language in chapter 5, and complex quantifier prefixes in chapter 6. We then took a *linguistic turn*[1] in chapter 7 by introducing *pragmatic* considerations in query processing, and completed the transition from queries to questions in chapter 8 by focusing in Question Answering. Throughout, the use of Generalized Quantification provided the glue that held all the material together as a coherent whole -hopefully!

Recall our thesis that querying is, essentially, a linguistic activity. We distinguished between *questions*, expressions in a natural language, and *queries*, expressions in a logical language -note that *logical language* is, in this context, a synonym for *declarative query language*. We claimed that past research has focused on either questions (the domain of Formal Linguistics, and Philosophy of Language) or in queries (the domain of Databases and Theoretical Computer Science). Both are indirectly related through their use of Logic as a tool; therefore, it should not be surprising that a logical concept (that of Generalized Quantifier) can be used to bridge the gap between fields. It is this potential to bring together insights from different disciplines, and the potential for *cross-illumination* (where concepts, techniques or tools from one discipline are used to promote growth in the other) that originated much of the research exposed here.

---

[1] With apologies to Richard Rorty ([86]). I always wanted to use this expression!

A.Badia, *Quantifiers in Action: Generalized Quantification in Query. Logical and Natural Languages*, Advances in Database Systems 37, DOI: 10.1007/978-0-387-09564-6_10, © Springer Science+Business Media, LLC 2009

However, much remains to be done in this respect, and the present work can only claim to have barely started. As stated in the Introduction, our of our main aims was to call the attention of researchers in one field to related work in another field. While it is clearly difficult to establish connections across disciplines, in the work presented here we have attempted to do exactly that. Hopefully, others will carry out further research on the themes sketched here: the surface has been only scratched, and the best fruits still await. Clearly, the last part of chapter 8 and all of chapter 9 offer several potential avenues of research, but other, brand new ones can probably be uncovered by researchers. For instance, research in Question Answering could profitable use the work on questions on formal linguistics ([20, 37]) to get a more formal footing on the subject. A more formalized framework will emphasize the similarities with querying, and then connections with database querying may in turn be developed. Only the future will tell if such connections are a profitable avenue of research.

Clearly, the reader must decide whether such enterprises are *worthwhile*. This book attempts to prove that they are *possible*, and argue that they are definitively *worthy* of further attention. In this sense, this book should be considered a snapshot of what is still work in progress, and an encouragement to others to establish collaborations and continue the work. In the end, this may result in somebody writing a book in the near future that surpasses this one in depth and coverage. We would consider this our greatest accomplishment.

References

1. The 9th text retrieval conference (trec 2000). online publication (`http://trec.nist.gov/pubs/trec9/t9_proceedings.html`). National Institute of Standards and Technology (NIST).
2. The tenth text retrieval conference (trec 2001). online publication (`http://trec.nist.gov/pubs/trec10/t10_proceedings.html`). National Institute of Standards and Technology (NIST).
3. The TPCH benchmark, `www.tpc.org`.
4. Fifty years of generalized quantifiers, conference in honour of professor andrzej mostowski. http://www.logika.uw.edu.pl/50yearsGQ/index.html, June–July 2007. Banach Center, Warsaw (Poland).
5. S. Abiteboul, P. Buneman, and D. Suciu. *Data on the Web: From Relations to Semistructured Data and XML*. Morgan Kaufman, 1999.
6. S. Abiteboul, R. Hull, and V. Vianu. *Foundations of Databases*. Addison-Wesley, 1995.
7. A. Badia. *A Family of Query Languages with Generalized Quantifiers: its Definition, Properties and Expressiveness*. PhD thesis, Indiana University, 1997.
8. A. Badia. Cooperative query answering with generalized quantifiers. *Intelligent and Cooperative Information Systems*, 1998.
9. A. Badia. Cooperative query answering with generalized quantifiers. *Journal of Intelligent Information Systems*, 12(1):75–97, 1999.
10. A. Badia. Safety, domain independence and generalized quantification. *Data and Knowledge Engineering*, 38(2):147–172, 2001.
11. A. Badia. Question answering and database querying: Bridging the gap with generalized quantification. *Journal of Applied Logic*, 5(1):3–19, 2007. special issue on Questions and Answers: Theoretical and Applied Perspectives, Rafaella Bernardi and Bonnie Webber, eds.
12. A. Badia, M. Gyssens, and D. Van Gucht. *Query Languages with Generalized Quantifiers*, chapter 11. Kluwer Academic Publisher, 1995.
13. A. Badia and S. Vansummeren. Non-linear prefixes in query languages. In *Proceedings of the Principles of Database Systems (PODS)*, 2007.
14. R. Baeza-Yates and B. Ribeiro-Neto. *Modern Information Retrieval*. Addison-Wesley, 1999.
15. J. Barwise. On branching quantifiers in english. *Journal of Philosophical Logic*, 8:47–80, 1979.

16. J. Barwise and R. Cooper. Generalized quantifiers and natural language. *Linguistic and Philosophy*, 4:159–219, 1981.

17. C. Beeri and Y. Kornatzky. A logical query language for hypertext systems. In N. Streitz, A. Rizk, and J. Andre, editors, *Hypertext: Concepts, Systems and Applications. Proceedings of the First European Conference on Hypertext.* Cambridge University Press, 1990.

18. R. Belew. *Finding Out About.* Cambridge University Press, 2008.

19. U. Chakravarthy, J. Minker, and J. Grant. Semantic query optimization: Additional constraints and control strategies. In L. Kerschberg, editor, *Proceedings from the First International Conference in Expert Database Systems.* 1987.

20. G. Chierchia. Questions with quantifiers. *Natural Language Semantics*, 1, 1993.

21. H. Christiansen, H.L. Larsen, and T. Andreasen. *Proceedings of the 1996 Workshop on Flexible Query Answering Systems.* Roskilde University Center, 1996.

22. W. Chu and Q. Chen. *a structured approach for cooperative query answering.* IEEE Transactions on Knowledge and Data Engineering, *1994.*

23. W. et alia Chu. *Cobase, a scalable and extendible cooperative information system.* Journal of Intelligent Information Systems, *6, 1996.*

24. *Anuj Dawar. Generalized quantifiers and logical reducibilities.* J. Log. Comput., *5(2):213–226, 1995.*

25. *Anuj Dawar and Erich Grädel. Generalized quantifiers and 0-1 laws. In* LICS '95: Proceedings of the 10th Annual IEEE Symposium on Logic in Computer Science, *page 54, Washington, DC, USA, 1995. IEEE Computer Society.*

26. *D. R. Dowty, R. E. Wall, and S. Peters. Introduction to Montague Semantics. Kluwer Academic Publishers, 1981.*

27. *H.-D. Ebbinghaus and J. Flum. Finite Model Theory. Springer, 2005.*

28. *H.D. Ebbinghaus. Extended logics: The general framework. In J. Barwise and S. Feferman, editors,* Model-Theoretic Logics. Springer-Verlag, 1975.

29. *H.D. Ebbinghaus, J. Flum, and W. Thomas.* Mathematical Logic. Springer Verlag, 1994.

30. *R. Fagin. Generalized first-order spectra and polynomial-time recognizable sets. In R. M. Karp, editor,* Complexity of Computation, volume 7 of SIAM-AMS Proceedings, pages 43–73. SIAM Press, 1974.

31. *T. Gaasterland. cooperative answering through controlled query relaxation. IEEE Expert, (5), September/October 1997.*

32. T. Gaasterland, P. Godfrey, and J. Minker. An overview of cooperative answering. *Journal of Inteligent Information Systems*, 1, 1992.

33. A. Gal and J. Minker. Informative and cooperative answers in databases using integrity constraints. In V. Dahl and P. Saint-Dizier, editors, *Natural Language Understanding and Logic Programming.* 1988.

34. R. A. Ganski and H. K. T. Wong. Optimization of nested sql queries revisited. In *Proceedings of the ACM SIGMOD Conference*, pages 23–33, 1987.

35. P. Gardenfors, editor. *Generalized Quantifiers.* Redeil Publishing Company, 1987.

36. Jean Mark Gawron and Stanley Peters. *Anaphora and quantification in situation semantics.* Center for the Study of Language and Information, 1990.

37. J. Ginzburg and I. Sag. *Interrogative Investigations: the form, meaning, and use of English Interrogatives.* CSLI Publications, Stanford, distributed by University of Chicago Press, 2001.

38. P. Grice. Logic and conversation. In P. Cole and J. Morgan, editors, *Syntax and Semantics*, volume 3. Academic Press, 1975.

39. J. Groenendijk and M. Stokhof. Questions. In J. van Benthem and A. ter Meulen, editors, *Handbook of Logic and Language*. Elsevier, 1997.
40. S. Grumbach and C. Tollu. Query languages with counters. In *Proceedings of the ICDT Conference*, pages 124–139, 1992.
41. A. Gupta, V. Harinarayan, and D. Quass. Aggregate-query processing in data warehousing environments. In *Proceedings of the VLDB Conference*, pages 358–369, 1995.
42. J.J. Gutierrez Reixach. Questions and generalized quantifiers. Master's thesis, Universiy of California, Los Angeles, 1995.
43. P. Hajeck. Logics for data mining. In *Proceedings of the ISAI Workshop*, 1998.
44. H. Han and M. Kamber. *Data Mining: Concepts and Techniques*. Morgan Kaufmann, 2000.
45. J. Han, Y. Huang, and N. Cercone. Intelligent query answering by knowledge discovery techniques. *IEEE Transactions on Knowledge and Data Engineering*, 8, 1996.
46. S. M. Harabagiu, D. I. Moldovan, M. Pasca, M. Surdeanu, R. Mihalcea, R. Girju, V. Rus, F. Lactusu, P. Morarescu, and R. C. Bunescu. Answering complex, list and context questions with lcc's question-answering server. In Voorhees and Harman, editors, *The Tenth Text REtrieval Conference (TREC 2001)*. National Institute of Standards and Technology, 2001.
47. L. Hella. Logical hierarchies in PTIME. In *Proceedings of the 7th IEEE Symposium on Logic in Computer Science*, 1992.
48. L. Hella, K. Luosto, and J. Vaananen. The hierarchy theorem for generalized quantifiers. *Journal of Symbolic Logic*, 61(3):802–817, 1996.
49. J. Hintikka. Quantifiers vs. quantification theory. *Linguistic Inquiry*, 5(2):153–177, 1974.
50. E. Hovy, L. Gerber, U. Hermjakob, M. Junk, and C-Y Lin. Question answering in webclopedia. In *Proceedings of the The Ninth Text REtrieval Conference (TREC 9)*, NIST Special Publication 500-249, 2000. http://trec.nist.gov/pubs/trec9/t9_proceedings.html.
51. P. Y. Hsu. *Incorporating Context into Databases*. PhD thesis, UCLA, 1995.
52. P. Y. Hsu and D. S. Parker. Improving SQL with generalized quantifiers. In *Proceedings of the Tenth International Conference on Data Engineering (ICDE)*, 1995.
53. R. Hull and J. Su. Domain independence and the relational calculus. *Acta Informatica*, 1995.
54. T. Imielinsky. Intelligent query answering in rule-based system. In J. Minker, editor, *Foundations of Deductive Databases and Logic Programming*. Morgan Kaufman, 1988.
55. T. Imielinsky and R. Demolombe. *Nonstandard Queries and nonstandard answers*. Oxford University Press, 1994.
56. N. Immerman. *Descriptive Complexity*. Springer, 1998.
57. J.M. Janas. On the feasibility of informative answers. In H. Gallaire and J. Minker, editors, *Logic and Databases*. 1978.
58. J. Kaplan. Cooperative responses from a portable natural language database query system. In M. Brady and R. Brewick, editors, *Computational Models of Discourse*, pages 165–208. The MIT Press, 1983.
59. E. Keenan. Beyond the frege boundary. *Linguistics and Philosophy*, 15, 1992.
60. E. Keenan. Natural language, sortal reducibility and generalized quantifiers. *Journal of Symbolic Logic*, 58(1):314–325, 1993.

61. E. Keenan and L. Moss. Generalized quantifiers and the expressive power of natural language. In van Benthem and ter Meulen [94].
62. E. Keenan and D. Westerstahl. Generalized quantifiers in linguistics and logic. In van Benthem and ter Meulen [94].
63. J. Keisler and W. Walkoe. The diversity of quantifier prefixes. *Journal of Symbolic Logic*, 38:79–85, 1973.
64. M. Kifer. On safety, domain independence, and capturability of database queries. In *Proceedings of the Third International Conference on Data and Knowledge Bases*, 1988.
65. P. Kolaitis and J. Väänänen. Generalized quantifiers and pebble games on finite structures. In *Proceedings of the 7th IEEE Symposium on Logic in Computer Science*, pages 348–359, 1992.
66. M. Krynicki and M. Mostowski. *Quantifiers: logics, models and computation*, volume I, chapter Henkin Quantifiers, pages 193–262. Kluwer, 1995.
67. H.L. Larsen and T. Andreasen. *Proceedings of the 1994 Workshop on Flexible Query Answering Systems*. Roskilde University Center, 1994.
68. Dirk Leinders and Jan Van den Bussche. On the complexity of division and set joins in the relational algebra. In *Proceedings of the ACM PODS Conference*, pages 76–83, 2005.
69. L. Libkin. *Elements of Finite Model Theory*. Springer, 2004.
70. P. Lindstrom. First order predicate logic with generalized quantifiers. *Theoria*, 32, 1966.
71. Godehard Link. *Algebraic semantics in language and philosophy*. CSLI Publications, 1998.
72. N. Mamoulis. Efficient processing of joins on set-valued attributes. In *Proceedings of the ACM SIGMOD Conference*, pages 157–168, 2003.
73. C. Manning, P. Raghavan, and H. Schutze. *Introduction to Information Retrieval*. Cambridge University Press, 2008.
74. N. May, S. Helmer, and G. Moerkotte. Quantifiers in xquery. In *Proceedings International Conference on Web Information Systems Engineering (WISE)*, pages 313–316, 2003.
75. D. Moldovan, S. Harabagiu, R. Girju, P¿ Morarescu, F. Lacatusu, A. Noischi, A. Badulescu, and O. Bolohna. Lcc tools for question answering. In Voorhees and Harman, editors, *The Tenth Text REtrieval Conference (TREC 2001)*. National Institute of Standards and Technology, 2001.
76. A. Mostowski. On a generalization of quantifiers. *Fundamenta Mathematica*, 44, 1957.
77. A. Motro. Using constraints to provide intensional answers to relational queries. In *Proceedings of the Fifteenth Conference on Very Large Databases (VLDB)*, 1989.
78. I. S. Mumick, S. J. Finkelstein, Hamid Pirahesh, and Raghu Ramakrishnan. Magic is relevant. *SIGMOD Rec.*, 19(2):247–258, 1990.
79. Inderpal Singh Mumick, Sheldon J. Finkelstein, Hamid Pirahesh, and Raghu Ramakrishnan. Magic conditions. *ACM Transactions on Database Systems*, 21(1):107–155, 1996.
80. M. Otto. Generalized quantifiers for simple properties. In *Proceedings of 9th IEEE Symposium on Logic in Computer Science LICS '94*, pages 30–39, 1994.
81. M. Otto. *Bounded variable logics and counting – A study in finite models*, volume 9. Springer-Verlag, 1997. IX+183 pages.

82. J.J. Quantz. How to fit generalized quantifiers into terminological logics. In *10th European Conference on Artificial Intelligence*, 1992.

83. R. Ramakrishnan and J. Gerhke. *Database Management Systems*. McGraw-Hill, 3rd edition, 2007.

84. K. Ramasamy, J. Patel, J. Naughton, and R. Kaushik. Set containment joins: The good, the bad and the ugly. In *Proceedings of the VLDB Conference*, pages 351–362, 2000.

85. S. Rao, D. Van Gucht, and A. Badia. Providing better support for a class of decision support queries. In *Proceedings of the ACM SIGMOD Conference*, Montreal, Canada, June 1996.

86. R. Rorty, editor. *The Linguistic Turn: Essays in Philosophical Method*. University of Chicago Press, 1992. reprint from the original 1967 edition.

87. S. Harabagiu S. Narayanan. Question answering based on semantic structures. In *Proceedings of the International Conference on Computational Linguistics (COLING 2004)*, 2004.

88. R. Scha. Distributive, collective and cumulative quantification. In *Truth, Interpretation and Information; selected papers from the 3rd Amsterdam Colloquium*, pages 131–158, Dordrecht, Holland, 1984.

89. M. Spaan. Parallel quantification. In J. van der Does and J. van Eyck, editors, *Generalized Quantifier Theory and Applications*, CSLI Lecture Notes. CSLI, 1993.

90. D. Speelman. A natural language interface that uses generalized quantifiers. In *Lecture Notes in Computer Science; Lecture Notes in Artifical Intelligence*, 1994.

91. E. Thijsse. Counting quantifiers. In van Benthem and ter Meulen [94].

92. The eleventh text retrieval conference (trec 2002). online publication at `http://trec.nist.gov/pubs/trec11/t11_proceedings.html`. National Institute of Standards and Technology (NIST).

93. J. Vaananen. Unary quantifiers on finite models. *Journal of Logic, Language and Information*, 6(3), 1997.

94. J. van Benthem and A. ter Meulen, editors. *Generalized Quantifiers in Natural Language*. Foris Publications, 1985.

95. J. van Benthem and A. ter Meulen, editors. *Generalized Quantifiers in Natural Language*. Foris Publications, 1985.

96. Johan van Benthem. Questions about quantifiers. *Journal of Symbolic Logic*, 49:443–466, 1984.

97. Jan van Eijk. Generalized quantifiers and traditional logic. In van Benthem and ter Meulen [94].

98. Allen van Gelder and R.W. Topor. Safety and translation of relational calculus queries. *ACM Transactions on Database Systems*, 16, 1991.

99. E.M. Voorhees. Overview of the trec 2003 question answering track. In *The Twelfth Text Retrieval Conference*, NIST Special Publication: SP 500-255, 2003.

100. W. Walkoe. Finite partially-ordered quantification. *Journal of Symbolic Logic*, 35(4):535–555, 1970.

101. D Westerstahl. Branching generalized quantifiers and natural language. In P. Gardenfors, editor, *Generalized Quantifiers*. Redeil Publishing Company, 1987.

102. D. Westerstahl. Quantifiers in formal and natural languages. In D. Gabbay and F. Guenther, editors, *Handbook of Philosophical Logic*, volume IV, chapter IV. Reidel Publishing Company, 1989.

103. D. Westerstahl. *Quantifiers: logics, models and computation*, volume I, chapter Quantifiers in Natural Language: A Survey of Some Recent Work. Kluwer, 1995.

104. D Westerstahl. Relativization of quantifiers in finite models. In J. van der Does and J. van Eijck, editors, *Quantifiers, Logic and Languages*. CSLI Lecture Notes, 1996.

105. G. Wiederhold. *mediators in the architecture of future information systems*. IEEE Computer, *March 1992*.

106. *J. Zhou, P. Larson, J. Freytag, and W. Lehner. Efficient exploitation of similar subexpressions for query processing.* In *Proceedings of the ACM SIGMOD Conference, pages 533–544, 2007.*

107. *C. Zuzarte, H. Pirahesh, W. Ma, Q. Cheng, L. Liu, and K. Wong. Winmagic: Subquery elimination using window aggregation.* In *Proceedings of the ACM SIGMOD Conference, pages 652–656, 2003.*

Printed in the United States of America